当代建筑师系列

齐欣
QI XIN

齐欣建筑设计咨询有限公司　编著

中国建筑工业出版社

图书在版编目(CIP)数据

齐欣/齐欣建筑设计咨询有限公司编著.—北京：中国建筑工业出版社，2012.7
(当代建筑师系列)
ISBN 978-7-112-14278-1

Ⅰ.①齐… Ⅱ.①齐… Ⅲ.①建筑设计-作品集-中国-现代②建筑艺术-作品-评论-中国-现代 Ⅳ.①TU206②TU-862

中国版本图书馆CIP数据核字（2012）第085857号

整体策划：陆新之
责任编辑：徐 冉 刘 丹
责任设计：赵明霞
责任校对：刘梦然 刘 钰

感谢山东金晶科技股份有限公司大力支持

当代建筑师系列
齐欣
齐欣建筑设计咨询有限公司 编著
*
中国建筑工业出版社出版、发行（北京西郊百万庄）
各地新华书店、建筑书店经销
北京嘉泰利德公司制版
北京顺诚彩色印刷有限公司印刷
*
开本：965×1270毫米 1/16 印张：11 字数：308千字
2013年1月第一版 2013年1月第一次印刷
定价：98.00元
ISBN 978-7-112-14278-1
(22347)

版权所有 翻印必究
如有印装质量问题，可寄本社退换
（邮政编码100037）

目 录　Contents

齐欣印象	4	Portrait
国家会计学院	6	National Accounting Institute
商业街坊	18	Retail Block
管委会	28	Government Buildings
玉鸟流苏	40	Shopping Arcade
天城售楼处	52	Sales Office
江苏软件园	64	Software Park
似合院	74	Down Plaza
西溪会馆	84	Xixi Club
Y-1-28 办公大厦	98	Y-1-28 Office Tower
演艺中心	106	Performance Centre
文化中心	114	Culture Centre
神农坛	122	Yan Di Temple
20+10 酒店	128	20+10 Hotel
学生活动中心	136	Students Centre
地铁网控中心	144	Metro-Traffic Control Center
齐欣访谈	152	Interview
团队	159	Team
其他作品年表	160	Chronology of other Works
齐欣简介	174	Profile
致谢	175	Acknowledgement

齐欣印象

文／黄元炤

出生于1959年的齐欣在北京长大。他于1983年毕业于清华大学建筑系，随后去法国留学，1996年年底回国。1990年前后，齐欣开始在不同规模的境外设计事务所里实习与工作，积累了丰富与厚实的经验。

在法国巴黎建筑与城市规划设计院（SCAU）工作期间，齐欣练就一身基本功。比如平面，一画就是三年。基于在国内工科学校打下的基础，他所提出的结构建议解决了不少疑难问题，从而获得事务所的高度认可。当时，齐欣经常与一位美籍建筑师搭班，齐欣负责解决功能问题，而美籍建筑师则负责创意与造型。那位美国人出手就能画出优美的线条，设计十分飘逸。这一特质，对日后齐欣的设计形成了某种影响——他的作品中时常隐藏或浮现着某种随意、自由与飘逸，带有独特的优雅和魅力。姿态是放松的，氛围是恬淡的。

在福斯特事务所工作期间，齐欣参与了香港机场货运站的设计，并开始接触国内规划、办公和空港等大型公共建筑项目。诺曼·福斯特（Norman Forster）是晚期现代主义中有高技倾向的建筑师，作品中体现了材料与工艺、技术结合下产生的纯净与张力。从齐欣设计的北京国家会计学院的主楼中能感受到来自福斯特的影响。这是一个带有技术倾向的设计：近似椭圆形的平面既简洁又典雅，虚化的玻璃与金属构件相搭配，给人一种大气之感，同时又极富有诗意。而后面的生活区，则带有浓浓的法国现代建筑的浪漫气息。

福斯特对齐欣的影响还在于设计前做大量的基础研究，并在研究、推理与逻辑衍生的基础上寻求突破。例如北京的贝克特厂房，齐欣先做了大屋顶采光模式的研究，并将这一研究的成果与中国的院落、方窗相结合，屋顶天窗的雏形逐渐演绎成墙面的开窗形式。

福斯特强调整体性，无论项目大小，都将其视作一个物件来设计。这也许对齐欣影响最大。齐欣的设计中，哪怕建筑的体量很碎或很错落，却都带有很强的整体感。如武汉融科天城售楼处，他用一张铸铝网把原本掰成两块的物体蒙了起来，形成一个完整的建筑；再如良渚的玉鸟流苏，在保证每个商铺个性的同时，他用一个屋脊将所有商铺串了起来，形成一个延续的带状商业街；在北京奥林匹克公园的下沉广场里，他用一个连续的钢构件来表述中国传统建筑的预制装配体系；在杭州的西溪会所中，他则将一个抽象的图案覆到变化和错落的几何体上，构成一幅整体的图画。哪怕有时刻意先将大体量东西击碎，齐欣也不会忘记最终建筑的整体感，并能从容地收拾残局，将破碎的物体完形。

20世纪90年代中期回国后，齐欣先到高校任教，之后进入设计市场。

北京国家会计学院的设计为他带来了很高的声誉，让齐欣在国内建筑界的声名鹊起。2002年，齐欣成立了事务所，任董事长兼总建筑师，开始以个人的名义从业。但由于会计学院的设计并未完全摆脱福斯特的影响，所以齐欣对自己并不十分满意。他未必觉得福斯特不好，只是认为笼罩在别人的阴影下，无法实现自己。他开始关注如何在当今世界的建筑语汇与思潮中确立自己的位置，但又不愿意采用唐突的方式，为树立个性而不遗余力。因而，他顺应自然，顺水推舟，顺理成章。齐欣对新的追求是不间断的，不断创新的秉性使他每次去尝试不同的设计方向。

齐欣喜欢用讲故事的方式叙述项目的操作过程。他对生成（分析、研究、概念）或生产（条件、纠葛、实施）的阐述非常清楚，而表达方式又完全不同于程式化的表格与图面，非常具有个人特色。

对于每个项目，齐欣都会有一套深思熟虑的独到想法。切入点会是文化、城市、结构或表皮。在做武汉融科天城售楼处的设计时，齐欣启用了外皮的语汇；在北京的贝克特厂房中，齐欣引入中国元素；在南京的秦淮风情街和良渚的玉鸟流苏项目中，齐欣制造了一个生动的城市场所；在杭州西溪湿地会馆的设计中，齐欣营造出一种虚幻的现实；在北京奥林匹克公园的下沉广场中，红色的圆环呼应了奥运的象征，而广场信息柱的设计则为树状，与鸟巢和水立方两个仿生建筑呼应。

综观齐欣的设计，可以发现每个项目都有不同的切入点、不同的思考，并且非常注重逻辑，有自成一体的研究分析方法以及对设计的表达。在设计倾向上，齐欣从早期的关注技术转向关注舒适与自然。可你又不知道他下一步往哪儿去，又有什么引人制胜的招数。给不同项目以不同的回应，这，就是齐欣的路，齐欣的特色。他可以关注社会、城市、环境、材料或技术，但只要已经做过了，就会改变。这意味着他很有艺术家的原创个性，忠于作品，更忠于作品的变。变，是他所建立的建筑哲学观与价值体系的基础。而变，又可以不温不火，优雅自在。他说他没思想，但没思想，也是一种思想。让我们拭目以待齐欣下一次的惊艳出手。

Portrait

By Huang Yuanzhao

Qi Xin, born in 1959, grew up in Beijing. He graduated from Architecture Department of Tsinghua University in 1983 and attended further studies in France in 1984. After returning to China at the end of 1996, he began to work in various design firms, accumulating a vast amount of experience.

While employed at SCAU of France, Qi Xin honed his fundamental design skills, such as building plan drawings, which he pursued over the course of 3 years. The engineering training he received at the university in China enabled him to actively participate in discussions on structure matters helping to solve some challenging issues and gaining him praise from his colleagues in the firm. At that time, Qi Xin often collaborated with an American architect. Qi was responsible for functional issues, while the American architect was in charge of architectural creation. His colleague who could easily draw out beautiful sketches, allows Qi to gain further inspiration and insight, which was directly reflected in his future designs. As a result, in Qi's works you can always feel a sense of casualness, freedom and grace, elegance and charm, as well as something relaxing and peaceful.

While working in the Norman Foster office, Qi Xin participated in the Cargo Terminal design in Hong Kong which began his transition into larger facilities projects. Foster is famous for his high-tech orientation. His work presents the purity and tension generated from the combination of material and techniques. Foster's impact is evident in Qi's design of the main building of the Beijing National Accounting Institute. It's a technique-oriented design: the oval shape is simple and elegant; the matching of glass and metal creates an outstanding effect with a poetic sense at the same time. In contrast, the accommodation area at the rear is shrouded with the romantic atmosphere of French architecture.

Foster's impact can also be seen in Qi's initial study of a building's function prior to its design. This logical, solution-based approach through careful research drive the project into a global and sometimes unpredictable solution. For instance, in Qi's Beckett factory design, the study on the lighting for a large internal space drove him to a solution of a very special roof, combined with a Chinese courtyard layout. The square pattern of the roof, as a component, extended gradually to the cladding.

Unity may be the most important impact Foster had on Qi Xin, as Foster treats every single project as one piece of object, whether it is large or small. In Qi's work, every scheme presents a strong sense of unity, even if the building is fragmented or separate. For example, in the case of the Sales Office in Wuhan, he covered the two separated parts of the building by a piece of aluminum net. So is the Yuniaoliusu Shopping Arcade, he connected all the stores by one ridge, making a continuous pedestrian shopping street without affecting the stores' individuality. In the Kind of Courtyard project in Beijing Olympic Park, Qi conveyed the pre-fabricated member of a traditional Chinese building by employing a contiguous steel structure. In the XiXi Club House of Hangzhou, he made first a motif, then, covered the whole building with it, although the building has a moving and overlapping geometry.

Qi Xin returned to China at the end of 1990s where he first taught at the university before returning to his career in architecture.

It was the design of Beijing National Accounting Institute that first brought Qi national recognition. It helped to establish his reputation as one of China's finest architects. In 2002, Qi Xin started his own practice. As the design of Accounting Institute was still under Foster's impact, he was not fully satisfied with his individuality as an architect. Never denying Norman Foster as a great architect, Qi also believed it impossible to fulfill himself under other's influence. He hasn't force himself to create a style of his own but rather chooses that it comes naturally and without extra effort. This character trait allows him to broaden his range of architectural orientations.

Qi Xin explains his work as a story, knowing the exact details of each and every phase of his projects.

His design may be based on a consideration of project's environment, or some culture issue, or the city, or just the cladding system. For the Sales Office, he explored the building with its envelope; in the Beckett factory, he incorporated the local courtyard in the plan; for the Retail Block and the Shopping Arcade, he created a rich and vibrant urban space; with the XiXi Club House, he turns reality into an illusion; in the Kind of Courtyard project, the red circles metamorphose into the Olympic Game's logo and a Chinese lantern; the information columns designed for the Olympic Park in Beijing stand like trees as if engaged in dialogue with the Bird's Nest and the Water Cube building on the site, which have both an organic architectural expression.

Different philosophy, different value, different guideline, different interpretation for different subjects, context, clients or simply different use, but always logical with himself, that may best describe Qi Xin's design. From his early technical-oriented design, Qi Xin has turned his focus onto what feels most natural and comfortable. He tried various design orientations in his past and once completed moves on. However, you never know what his next step will be. Much like artist, Qi understands the importance of change and finding pleasure wherever it exists. But no matter how important change may be, his designs maintain a certain degree of moderation and grace. He claims that he does not have any design philosophy, which is exactly his philosophy. Let's see what surprises Qi Xin will show us next.

国家会计学院 北京
National Accounting Institute, Beijing
1998 ~ 2001

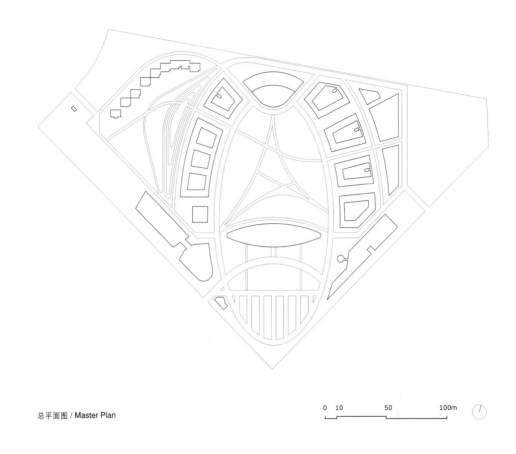

总平面图 / Master Plan

　　国家会计学院被定位为"经济建设时期的黄埔军校"。学院规模为1500名在校学员，培训期为三个星期至三个月，占地面积200亩，建筑面积7万平方米，建筑限高18米。

　　北京的机场高速路使顺义天竺开发区的所有路径均旋转了45°，与北京城的路网格局大相径庭。规划将此地块内的轴线重新转回到正南正北方向，坐北朝南。而避免轴线扭转所带来重叠网格冲突的最佳几何形体莫过于圆。因此，地块的中央呈现了一个椭圆，其中央偏南摆下了教学主楼，并以此为界，分出了前部的教学区和后部的生活区。

　　主楼面对的是一片林荫停车场，并被左边一个学生活动中心，右边一个图书馆相持。这两幢建筑均沿街而置，断面上呈四分之一圆，高度均由低向高发展，以谋求与未来周边的别墅建筑在高度上的协调。

　　教学主楼的后方为被一组学员公寓围合成的马蹄形绿地。每幢学生公寓均围绕内院组织，并由南侧的3层楼高逐渐过渡到北端的6层，以争取最大面积的日照。每个内院还拥有自己的色彩，它们的色彩将与院内的种植花色相配，形成鲜明的个性。七个院中不同的色彩还星星点点地反映在了围绕中心绿地的立面阳台上，形成一部明快而丰富的交响曲。

　　学员宿舍最北端建筑的底部为学生食堂。下课后的学员经由中心绿地来食堂就餐，加之联系东西两块户外活动场地的蜿蜒的纽带，若干路径将绿地切成了若干块。这些碎块就像是被敲碎的蛋壳，各自向不同的方向起翘，翘起部位的挡土墙内暗藏光带，从单侧照亮小路。植物的配置进一步强化了"蛋壳"的倾斜体量，它们从乔木过渡到灌木，再过渡到花草。每一片"蛋壳"上的花草均由各自不同色系的花果构成，春来冬至，它们将不断地涌现出独唱、重唱与合唱。

　　学员宿舍的东侧为后勤楼和体育馆，它们与宿舍的东立面一起界定了一条在国内不多见的城市型道路。而学员宿舍的西侧为一座人造山丘，其西侧还有一汪湖水和沿湖建造的专家公寓。

　　主楼前其实也有一个小坡地，它由北向南升高。站在主楼前，你将看到坡地上的草皮花卉和坡后的树冠，而看不见林荫停车场上的车辆。

　　整座学院的规划从城市设计入手，照顾与周边道路的关系，用建筑围合空间。建筑设计力求简洁，与功能紧密结合，并努力开发新的建材与技术。

功能	学校
规模	70000m²
阶段	竣工
甲方	财政部
乙方	京澳凯芬斯
团队	齐欣 + 高银坤 + 张文锋 + 沈立众 + 韩鉴

Function	School
Scale	70000m²
Phase	Bulit
Client	Ministry of Finance
Design	Jin-Ao Kann Finch
Team	Qi Xin + Gao Yinkun + Zhang Wenfeng + Shen Lizhong + Han Yin

The master plan tries to reply to the surrounding roads which are not north to south oriented like in Beijing city center. But inside of the campus, the main axe is returned back to a north to south orientation for a better sunlight. The buildings are organized around the central garden in an ellipse shape. The main building is located in its center, which separates the education buildings and the residential.

The architecture is resolutely modern, trying to meet the functional requirements, and implementing new building materials and technologies.

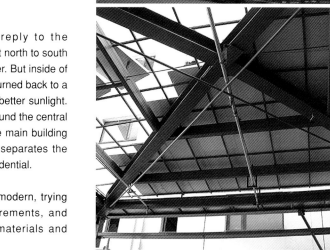

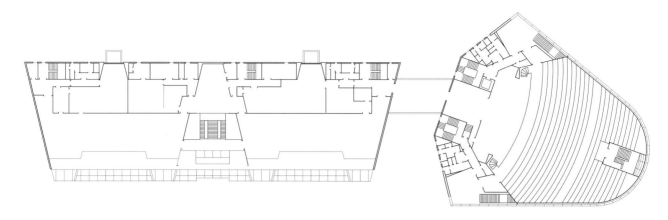

平面图 / Plan

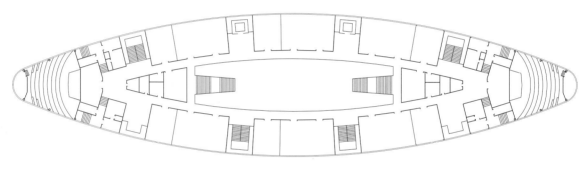

平面图 / Plan

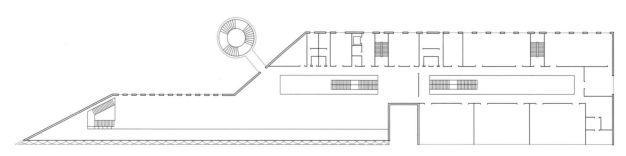

平面图 / Plan

0　5　10　20m

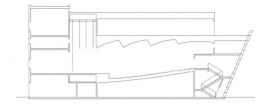

剖面图 / Section

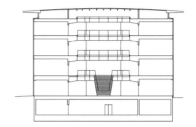

剖面图 / Section

剖面图 / Section

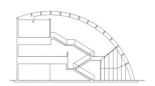

剖面图 / Section

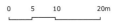

摄影：方振宁

摄影：方振宁

摄影：方振宁

摄影：付兴

摄影／付兴

摄影：方振宁

摄影：齐欣

摄影：付兴

商业街坊 廊坊

Retail Block, Langfang
2000 ~ 2002

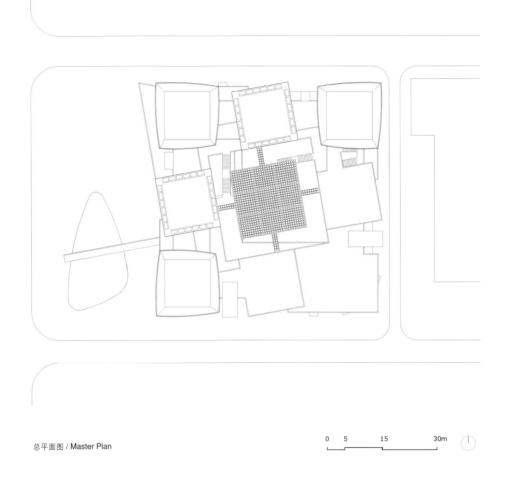

总平面图 / Master Plan

廊坊在老城区和北部新开发的区域间规划了一条贯穿东西的商业街，由东向西依次命名为第一、二、三、四、五大街，接下去还要建第六、七、八大街。街道不宽，却拥有60米纵深的街区。我们介入设计时，除第五大街尽西头那块60米×60米的地还空着外，其余的建筑均已照欧式风格落成，洋洋洒洒，色彩斑斓，颇为壮观。

开发商最初想象的这栋建筑也许是三张60米×60米摆起来的大饼，但这与销售经验相矛盾——因为只有临街的店铺才好卖。通过沟通，我们所达到的共识是：

- 第五大街的商业街区有着自圆其说的步行系统，应贯穿始终。
- 应借助街区南侧的城市公园这一资源，并与其对话。
- 作为第五大街的收头建筑，在体量上应与城市干道银河路两侧的新建筑相匹配，同时又要兼顾到基地东侧的小房子。
- 它不仅是第五大街的结束，也是第六大街的开始。古典建筑的语汇可以从此消失，转向现实的、有朝气的、面向未来的现实世界。

为了让所有的商铺都沿街，我们在房子的横竖方向各开出两条道。为了节省造价，中庭上方没有顶，实际成了个院子。为了把顾客引上五楼，院子里有穿插着的扶梯、楼梯和电梯。院子当中还支了个大棚子，棚子顶上会有乐师演奏。光看到乐师的脚当然不过瘾，那就上一层楼。看见了腿，还不过瘾，那再上。不知不觉地爬完四楼，上到五楼，终于，看到了乐师的天灵盖。为了强化与侧后方城市绿地的关系，插在中间的四个盒子转了个小角度。忽然，沿银河路一侧的主立面也迎向了老城的中心。为了使建筑的外立面尽可能地得到控制，作为店铺背面陈列货架的部位被墙体封实，而店铺的背立面恰恰是展示给城市的建筑正立面。为了强调每家店铺的个性，并使建筑与第五大街已有的色彩斑斓的建筑相匹配，围着院子的四个房子色彩各异，而把角的房子则一律呈白色，以统辖建筑的整体。

既然是商业建筑，就有点花，有点乱，这也是对齐欣一向偏于理性设计方式的一种挑战。

国内许多城市中都有所谓的"欧式"建筑或街区，大家见怪不怪。看到这么一栋彻底现代的建筑，人们仿佛见到了天外来客。有意思的是，当一位从巴黎蓬皮杜中心的策展人去看这组房子时，她却对那些仿古的建筑极为稀罕，觉得落入了一个难以置信的超现实国度。当她看到商业街坊这栋建筑时，才舒了一口气："终于回到了我们这一时代的现实之中。"

功能　　商业
规模　　13000m²
阶段　　竣工
甲方　　华夏地产
乙方　　京澳凯芬斯 + 维思平
团队　　齐欣 + 朱锦巍 + 吴钢 + 张瑛

Function　Retail
Scale　　13000m²
Phase　　Bulit
Client　　L.F. Chinese Property
Design　　Jin-Ao Kann Finch + WSP
Team　　Qi Xin + Zhu Jinwei + Wu Gang + Zhang Ying

Four of eight cubes rotate slightly to match the park situated at the back of the buildings. The building as a whole is in disorder because of its shopping and leisure use.

Many European style buildings have appeared all over China, so that local people take this buildings for granted. Something totally new or contemporary is suddenly beyond their expectations. Meanwhile, when a curator from Pompidou Center came to visit this building, she was rather surprised by seeing so many European style buildings on its side: this is just unbelievable! Only when she arrived at the Retail Block at the very end of a very long street, she felt relieved: we finally came back to a real world!

平面图 / Plan

0 5 10 20m

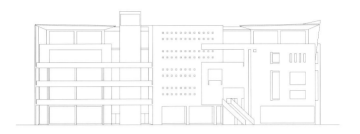

立面图 / Elevation

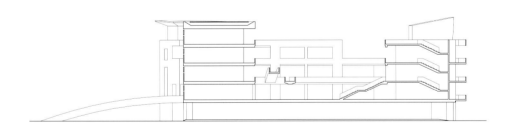

剖面图 / Section

摄影：方振宁

摄影：齐欣

摄影：方振宁

管委会　　东莞

Government Buildings, Dongguan
2002～2005

总平面图 / Master Plan

　　管委会办公小区是整座园区的行政中心，位于松山湖的北岸，呈半岛形。地块除去西南两个临水的界面外，北临一条过境快速路，东接城市礼宾林荫道。林荫道是一条贯穿南北的景观轴线，在地块以北串联着若干文化建筑，并在地块东翼形成末端，端部为供市民休闲所用的广场。跨林荫道地段以东正在规划一组会议中心及五星级酒店。

　　地块本身已经过平整，与周边起伏的丘陵地形形成对比。任务书中所要求的建筑面积并不很多，内容相对单纯（基本为办公），因此在规划中如何处理好小区与周边环境的关系便成为我们首要关注的问题。

　　为了充分利用地块，我们首先将办公楼定义为群体建筑，将原本相对集中的两个办公区（管委会办公大楼和小型独立办公楼）打散后重新组合，围合出一片宽阔的中心绿地。这种规划一方面能使所有的办公建筑均享有较好的日照、通风和景观条件，同时不对湖景构成压力。

　　地块中需要着重处理的建筑立面分别为沿礼宾大道的东立面和沿湖岸的南立面。规划中我们将小型独立办公楼以及一个展览中心布置在东侧，以便于他们的对外办公；将管委会办公楼布置在南面，使这些楼一方面能够充分地享受湖景，同时还拥有一个安静的绿化庭院。

　　为使管委会建筑更充分地享有湖色和南来的光线，并使其进一步与周边环境完美地结合，我们将这些小楼的体形塑造成三角形：双面临湖朝南，单面朝北，面向集中绿地。建筑从绿地一侧逐渐高攀，从而在重新找回被夷为平地的丘陵地形的同时塑造了判若船体的沿湖建筑形象。与之相应，小型独立办公楼在礼宾大道一侧也呈坡状，舒缓的草坡连接着继续高攀的藤萝架，架下为人行道和停车空间。由此，将景观与建筑紧密地结合在一起。

　　三角形的管委会建筑有五幢，分两种，共同环抱着中心绿地。大的一种只有一幢，占据着半岛上凸向湖面的地块，其内部有一共享大厅，是这组建筑的主楼。小的一种有四个，内部呈退台。这些建筑通过地下的一个环状车库使管委会办公楼形成了一个整体。建筑的外墙材料原设想采用木纹铝板，经甲方要求，调整为涂料。屋顶由金属和玻璃构成。

　　小型独立办公楼和展览中心体形一致，排成一排，四幢建筑在西侧由玻璃通廊相连，东侧则有布置在楼宇之间的雨廊。每幢建筑内部各有一个4层通高的中庭，面向中心绿地。

功能	办公
规模	42000m²
阶段	竣工
甲方	松山湖科技产业园管委会
乙方	齐欣建筑 + 北建工设计院
团队	齐欣 + 朱锦巍 + 赵占北 + 桂琳

Function	Office
Scale	42000m²
Phase	Bulit
Client	S.S.H. New Tech. Park
Design	Qi Xin Archi. + BJADARI
Team	Qi Xin + Zhu Jinwei + Zhao Zhanbei+ Gui Lin

When the project was handed to us, unfortunately, the site had already been leveled, which meant that the site was flat, in contrast with the undulating surroundings.

Was it possible to reverse the situation?

Along the front of the lake, five buildings were designed in a triangular shape, benefiting the lake views and the sunshine, and in harmony with the surroundings. They gradually rise from the central park. From the park side, they look like hills, whilst, from the lake-side, and they give an impression of boats. Correspondingly, smaller buildings are slopping along the urban avenue. The grass slope links four buildings with a continuous structure, over a sidewalk and a parking lot.

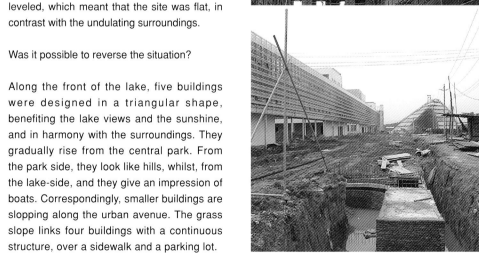

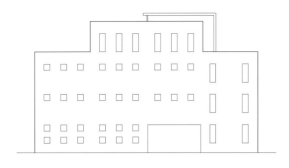

立面图 / Elevation

剖面图 / Section

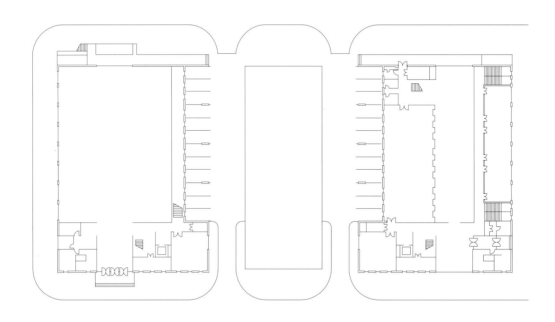

平面图 / Plan

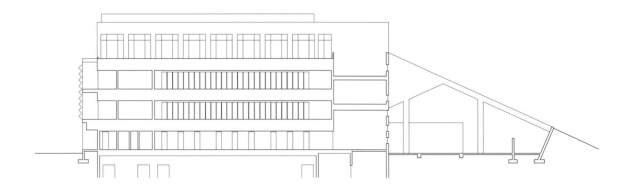

剖面图 / Section

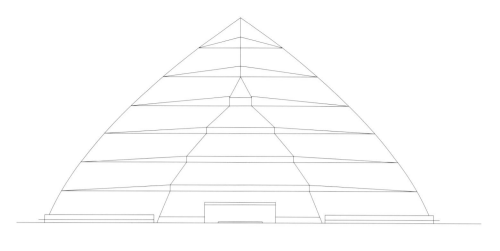

立面图 / Elevation

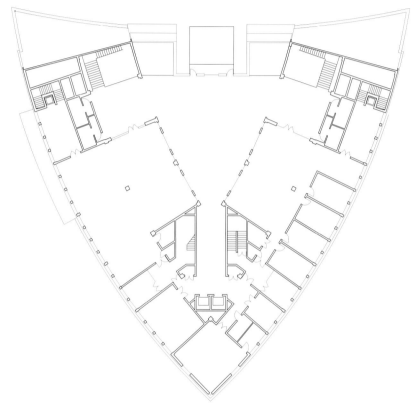

平面图 / Plan

剖面图 / Section

0 1　　　　　10　　　　　20m

摄影：齐欣

摄影：齐欣

摄影：齐欣

摄影：齐欣

摄影：齐欣

摄影：齐欣

摄影:齐欣

摄影：齐欣

玉鸟流苏 杭州
Shopping Arcade, Hangzhou
2004 ~ 2008

总平面图 / Master Plan

玉鸟流苏应当算是个旅游项目，内容基本是吃喝玩乐，是良渚文化村的一部分，紧挨着杭宁高速。这里原来就有个村庄，地势起伏。参与设计的四个事务所有张雷工作室、大舍、维思平和我们。张雷和我们的设计盖好了一半，其他的还在施工。

在我们这个设计团队介入之前，一个加拿大的景观设计公司做了规划，样式完全是北美的：一群孤独的房屋散布在田野中，找不着街道，找不着广场。我们的介入先从对这一规划的批判开始，大家一致认为不仅要呈现自然村落的形态，还要掌握好公共空间的收放。大舍为各家的工作做了分工，大家又一起商讨出一系列"游戏规则"，诸如檐口高度为8米，屋顶的倾角为30°，色彩以黑白灰为主，局部也可用木材等。

与其他几个地块相比，我们负责设计的地段进深局促，以线性延展。线性体的一侧为步行街或广场，另一侧面向机动车通道，是物资进出的服务性界面。

建筑的功能大致分三类：餐饮、商铺和艺术家工作室。餐饮、店铺始终沿街，而艺术家工作室却相对独立，享有一片世外桃源。

设计首要考虑的是如何将原地形中蜿蜒的街巷和生动的房屋挽留在这片土地上。

顺着狭长的地块，我们拉出了一条长长的飘带，延续的屋脊贯穿始终。带状建筑的走向限定了街道的宽窄和广场的形态。

带下的商铺是"巨构建筑"中的活力要素，它回顾着原始村落的随意，彰显着各路商家的特色。店铺单元的个性释放表现在同一屋脊下的前后错动和左右翻转，并不经意地挤压出异样的交通空间。

在千姿百态、表情各异的店铺门脸和街面之间，一条长长的骑楼伺候着企图遮阳避雨而又无所事事的游客。骑楼外侧，时疏时紧的柱廊似乎仍在羞涩地延续着活跃的气氛，但纤细的柱廊，仿佛一排琴弦，已悄然将后部一段段璀璨斑斓的乐句梳理成了一部协调而动听的乐章，并规范了广告和招幌的位置。

白粉墙、灰瓦顶再一次将个性收拢，散发出清淡的地域情调。鱼鳞状的灰瓦一直蔓延到建筑的背后，张合有致，伴随着建筑的呼吸。

线性建筑随着街道汇聚到村中央时，一个形态简单而奇特的建筑在此画下惊叹号。它判若村落中的祠堂，聚焦着停顿人群的目光。

院落在两个相互倒置的艺术家工作室中出现，并伴随着呈L状的展廊。光线从展室围墙上端的缝隙中徐徐地飘进室内，塑造出柔和而宁静的氛围。

功能	商业
规模	8500m²
阶段	竣工
甲方	浙江万科
乙方	齐欣建筑 + 京澳凯芬斯
团队	齐欣 + 刘尔东 + 高银坤 + 张亚娟 + 王斌 + 罗斌

Function	Retail
Scale	8500m²
Phase	Bulit
Client	Zhejiang Vanke
Design	Qi Xin Archi. + Jin-Ao Kann Finch
Design	Qi Xin + Liu Erdong + Gao Yinkui + Zhang Yajuan + Wang Bin + Luo Bin

Four teams work together to design this tourist project. As one team, the given site is both narrow and long. The priority for the team is to keep the memory of the street pattern and the barn houses typology that used to belong to this site.

A continuous long roof passes through the whole site, which determines the shopping street and the square. Under the mega long roof, the individuality of each store is displayed by the overlapping and the rolling-over of every adjacent unit.

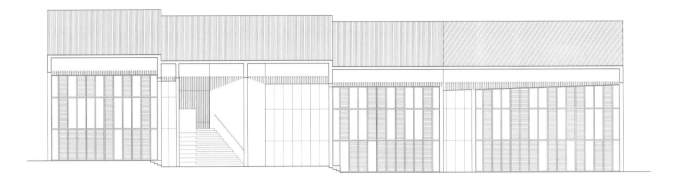

立面图 / Elevation

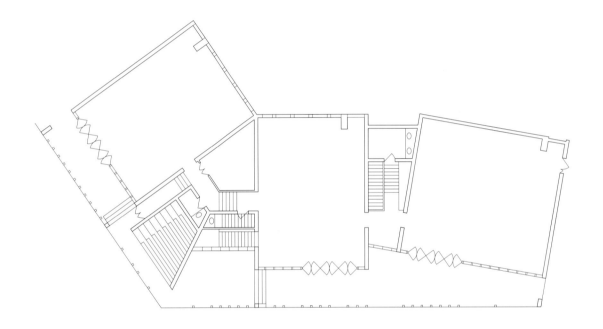

平面图 / Plan

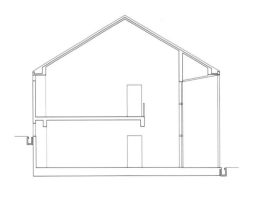

剖面图 / Section

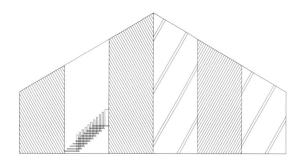

立面图 / Elevation

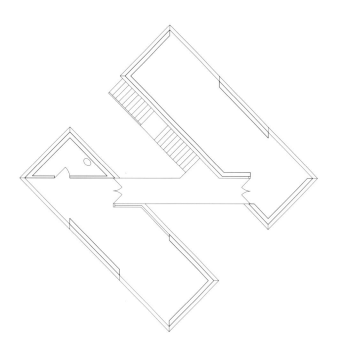

平面图 / Plan

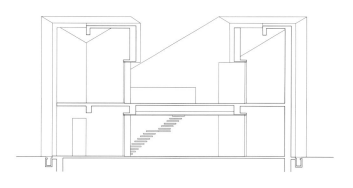

剖面图 / Section

摄影：杨超英

摄影：杨超英

摄影：杨超英

摄影：杨超英

摄影：杨超英

摄影 杨超英

摄影：杨超英

摄影：杨超英

天城售楼处 武汉
Sales Office, Wuhan
2006～2007

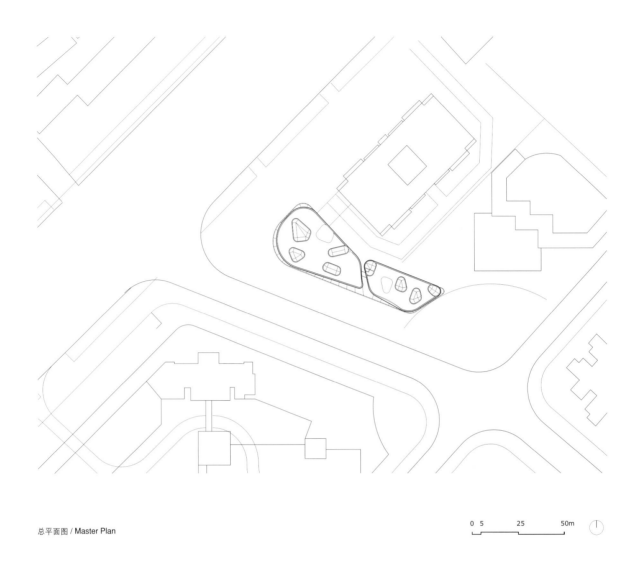

总平面图 / Master Plan

这是一个为配合高层住宅区而设的售楼处。售楼功能消失后，它将变成商业建筑，与其他一些小建筑共同围合出一个内向型的广场。

既然是售楼处，就要抢眼，以起到广告作用；既然被高大的住宅楼所包围，暴露在众目睽睽之下的屋顶就很重要。

从"抢眼"的角度出发，方案试图把建筑的体量做大，这就要把不多的空间整合。建筑的外轮廓线完全出自场地的用地红线：两条道路相交后抹出的圆角直接成了建筑的外界面，这一圆角相继延伸到建筑的各个角落。被抹了角的立面忽然变长，长长的立面为建筑营造了声势，造就了个性。

当时，场地上有一家钉子户，不知能不能拆，所以要考虑分期实施。为此，建筑被分成两段：一部分承担销售和办公的功能，可先建；待建的部分做样板间。显然，出自整合空间的目的，被一分为二的建筑还需再合二而一。

售楼处或商业建筑都需要引人入胜的室内空间。方案先将建筑的内部捅了几个大窟窿，接着又在屋顶上捅了几个小窟窿。大窟窿是一些公共活动场所；小窟窿是一些形态各异、自由分布的天窗，活跃着屋顶的表情。窗洞下铺设了色彩各异的丝绸，天光经过打了折的丝绸带着柔和的色彩均匀地到达室内，营造出不同的视觉观感。

还从塑造体量和生动室内的角度出发，方案将建筑分离出若干张皮。最里边的一层实际是核，包裹着楼梯间，最实。外面一圈是建筑的维护，它面向城市的一侧通透，面向后街的部分封闭。通透的部位还被分出透明与不透明两种材质，封闭的部位只在必要的位置开了些小窗。最后一张皮是行头，虚无缥缈。这张纱幕在最需要室内外交流的部位被剪开，让立面流动起来，与平面的曲线相得益彰。镂空的铸铝网格一方面在向城市彰显着自己的特色，还过滤着夏日强烈的光线。白天，日光投下的阴影丰富了室内的环境；入夜，灯光又把生动的阴影洒向街面。洒出的冰裂纹图案与住宅区的景观主题相吻合，或多或少地描绘着未来居民的生活场景。

功能	售楼处
规模	1400m²
阶段	竣工
甲方	融科智地
乙方	齐欣建筑 + 武汉市建筑设计院
团队	徐丹 + 齐欣 + 刘阳 + 张亚娟 + 王斌

Function	Sales Office
Scale	1400m²
Phase	Bulit
Client	Raycom Real Estate
Design	Qi Xin Archi. + WIAD
Team	Xu Dan + Qi Xin + Liu Yang + Zhang Yajuan + Wang Bin

This Sales Office is built for commercializing tower buildings of a new developed residential quarter. For this reason, the roof and the façade of this building need to be attractive.

Two distinct parts of the building are covered by one single skin. The envelope composed of prefabricated cast-in aluminum panels is a shade for the internal space, which creates an illusionary indoor sensation. In addition, the dramatic effect is emphasized by the light coming from the openings on the roof, where silk material of different colors is used to filter the sunlight, offering a colorful atmosphere.

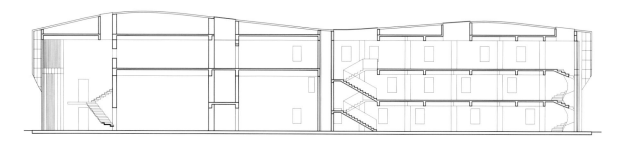

剖面图 / Section

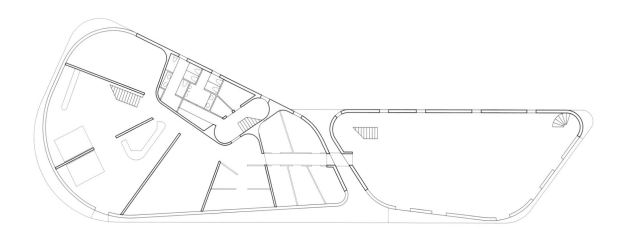

平面图 / Plan

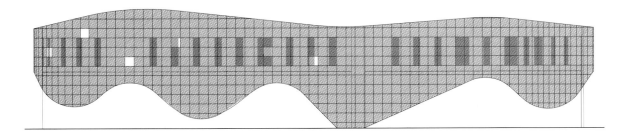

立面图 / Elevation

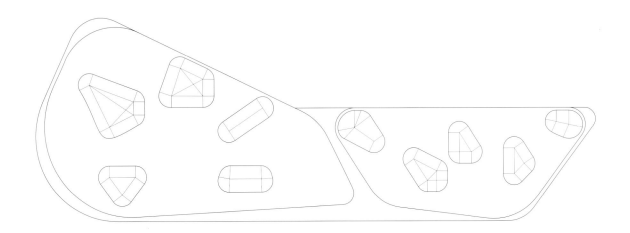

屋顶平面图 / Roof plan

摄影·齐欣

摄影 | 齐欣

摄影：齐欣

摄影：齐欣

摄影：齐欣

摄影：齐欣

江苏软件园 南京
Software Park, Nanjing
2006～2008

总平面图 / Master Plan

江苏软件园由诸多小办公楼及其配套公建组成，由张雷工作室、大舍和我们三家共同设计。项目处在南京的江宁新区，甲方是南京新城股份公司，策划是北京的华高莱斯，总图规划由张雷完成。"分地"的原则为各家任务基本相当，用"抓阄"的形式确定各家的具体地块。在分配中，张雷为每家留下了一块"高地"，建筑应因地制宜。关于建筑形态，华高莱斯与甲方制定了"新中式"的方向，提出"将院落进行到底"。有趣的是，当三家将各自的模型拼到一起时，所有人都为其整体感之强而感叹。其原因不仅是甲方与设计团队在设计前有过充分的交流，更是因为负责总体规划的张雷在一开始就编制了一套行之有效设计导则。

我们负责设计了三种类型的办公建筑，外加一组单身公寓和托儿所以及一个食堂。

第一组办公建筑以"点"的姿态站在空地间。"点"的形状是一个立方体，其中埋伏了十来个上上下下、大大小小的院子。建筑基本被竹廉遮了起来，只有院落处罩着花棱窗。另有一条纤细的缆桥在空中将院落与旁边高起的道路相连。

第二组办公建筑散布在高地上，为了减弱对丘陵的压迫，设计将面积本不很大的建筑分成三段：行人从二层的平台上进入室内，先要穿越一条铺架在一、二层共享空间中的小桥，才来到接待区；从接待区下一层，有被地势和建筑共同围拢的院子，其中的一部分有顶，成为户外的起居空间；若从二层上行，三层不仅有着视野开阔的外廊，还有相对私密的小院，院墙上开着为观望周边景观而设的窗洞。

第三组办公建筑靠着城市道路，外侧的地势低于内侧。建筑被分成一个站着和一个躺着的两部分。从内侧街道可直接登上躺着的建筑之屋顶花园，望远方的山景，或从夹缝进入下沉的庭院，再由此分别进入躺着或站着的两个建筑。当然，无论在哪个建筑中你都会找到"进行到底"了的院落。单身公寓是一组55平方米的跃层小住宅，一侧临街，一侧面对相对陡峭的山体。住宅呈带状环绕着托儿所。托儿所的屋顶上冒出大大小小的圆环，反映着下部的院落或天光。时而，也会有树木从环中穿出。在这片宁静的盆地中，一系列小桥插向公寓的楼体，成为每一住户的入口。

食堂由两个圆滑的几何体组成，外面绕了一圈围墙。建筑的外墙完全通透，而围墙却半隐半透，带来了一丝含蓄，一缕悬念。

功能	办公
规模	8300m²
阶段	竣工
甲方	南京新城发展股份有限公司
乙方	齐欣建筑 + 京澳凯芬斯
团队	齐欣 + 刘尔东 + 竹丽雅 + 王斌 + 高银坤 + 张亚娟

Function	Office
Scale	8300m²
Phase	Bulit
Client	NJXC Development Co.Ltd
Design	Qi Xin Archi. + Jin-Ao Kann Finch
Team	Qi Xin + Liu Erdong + Julia Verspieren + Wang Bin + Gao Yinkun + Zhang Yajuan

The client demands to design Chinese courtyards everywhere in the office building.

The first type of building is a compact block, in which ten courtyards spread at different levels; the second one lays on a varied topography, and its courtyards are at times standing on the ground on the lowest level, and at times at the top of the building, where more privacy is needed; the third type of building is a variation, with a different management and volume. Whilst the first type of building has a bamboo cladding system, the second and third one have just a simple black or white traditional wall. In addition, there is also a dormitory and a small canteen that has an architectural motif inscribed on a hollowed enclosure.

The enclosure, together with the black and white colored wall, bamboo cladding, courtyard, refers obviously to the local traditional housing typology.

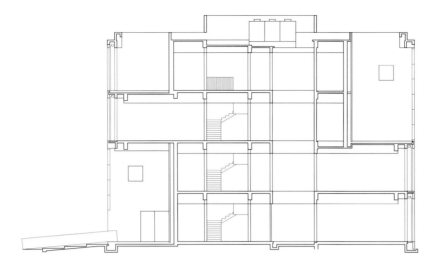

剖面图 / Section

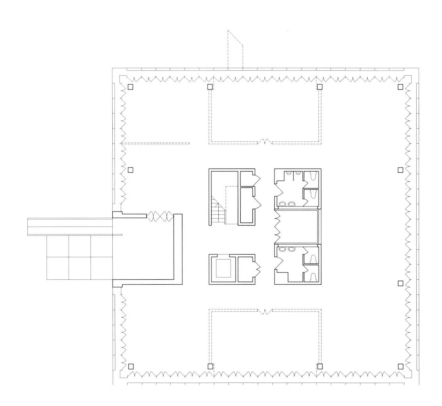

平面图 / Plan

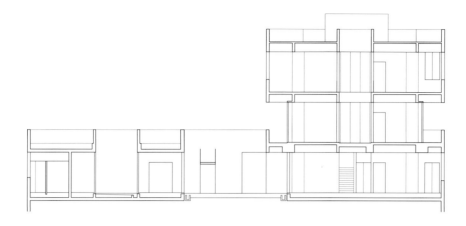

剖面图 / Section

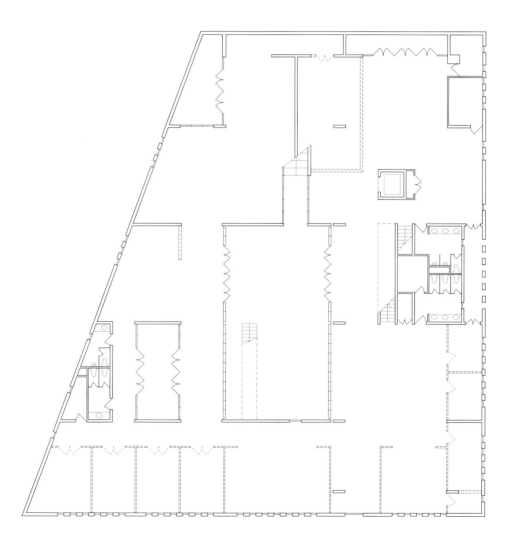

平面图 / Plan

摄影 舒赫

摄影：舒赫

摄影：舒赫

摄影 舒赫

摄影：舒赫

摄影：舒赫

似合院 北京
Down Plaza, Beijing
2007 ~ 2008

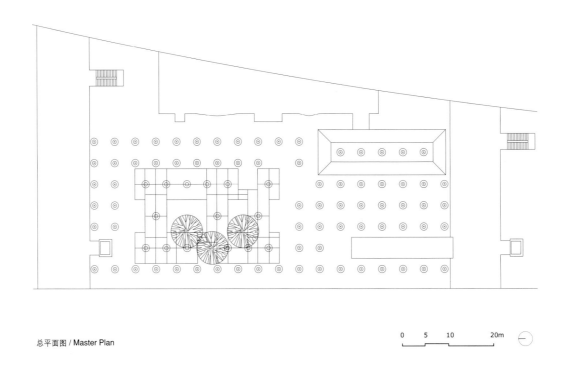

总平面图 / Master Plan

将历史与现实、全球与地域相混合后生成某种新的时空文化，是这一设计的初衷。

四合院群落的屋顶是老北京城的一大特色。而这片下沉空间比公园场地的标高低出了9米，使展示建筑的第五立面成为了一种可能和必要。

除了屋顶，作为老北京城的基本构成要素，四合院这一建筑类型还具有如下特点：

- 相对于城市，它是封闭而私密的居住场所，空间组织中充分体现了外合内敞的原则；
- 建筑群体是由一系列简单矩形平面构成的"间"相拼而成的；
- 其结构为梁柱体系，或曰框架体系，具有所谓的"墙倒屋不塌"的特征；
- 其建造元素完全由预制构件组成。

私密与围合显然与这一公共场所的性质相左。而这种矛盾和对立恰恰促成了从另一视角观赏中国建筑的契机。利用"墙倒屋不塌"的原理，方案清除了所有的围合物，以彻底打破室内外的界线，从而将原本封闭和私密的空间转换成开敞而公共的场所，让人们得以在无拘无束的自由行走过程中，既能品味似曾相识的感觉，更能体验到一种新生。

在一个被9米柱网围合的限定空间中，有另一层4.5米的柱网统辖了这里的景观。柱网上的一些柱子打了个弯，与梁连体，其上支檩，檩上架椽，椽上铺瓦，与传统四合院建筑一脉相承。构架中依循祖先的做法广泛应用预制构件，只是材料不再是木头，而是钢。钢的特性使原来相对繁琐的造型被简化，反映出时代的气息。

由梁柱支撑的遮阳棚为下沉广场两侧的商业、餐饮创造了在户外拓展的条件。其中的一条回廊在条凳的伴随下围合出一片水面，浅浅的水池中铺满了晶莹的玻璃珠，冬季可在视觉上取代水体。

四合院坡屋顶的造型在西侧的商业建筑立面上得到演绎。仍然是同一规格的钢管，还保持着同样的坡度，但在这里却被竖向搁置，构成具有韵律感的墙面和出入口。

为使人们在不同的高度上观赏到不同的景色，方案利用9米的下沉高度，将景观设计从简单的二维空间延伸到三维：柱网上的大部分柱子直冲云霄，上面悬挂着高高低低的红色圆环。圆环在呼应了奥运标志的同时，依据高度的变换，时为座椅，时为台面，时为灯笼。入夜，红色的圆环将营造出缥缈梦幻的景象。

灯笼的原型在东侧的商业建筑立面上得到了另一种演绎，这时它更贴近"笼"，而非灯。当然也是灯，那是在月亮接替了太阳的位置以后。

功能	景观
规模	3300m²
阶段	竣工
甲方	新奥集团
乙方	齐欣建筑 + 北京市建筑设计研究院
团队	齐欣 + 张亚娟 + 徐丹 + 王斌

Function	Landscape
Scale	3300m²
Phase	Bulit
Client	Xin Ao Group
Design	Qi Xin Archi. + BIAD
Team	Qi Xin + Zhang Yajuan + Xu Dan + Wang Bin

Inspired from the traditional courtyard house, this project provides a rich architectural bird-view expression.

The shelter constructions follow the traditional courtyard house model, using the same column-beam structure and the prefabricated assembly system. The concept in traditional Chinese architecture such as the "walls fall but not the houses" is displayed in a way that the walls disappear, and the simple steel elements replaces the columns, which on the one hand reach the roof, and on the other hand, cover the retail facilities, and become the only structure. Therefore, the project transforms the traditional courtyard house from an introverted domestic architecture into an open public space. A luminous circular element hanged on the dense masts at different levels, fills the whole space, symbolizing the Chinese traditional lanterns which float on the sky.

摄影／舒赫

摄影：齐欣

摄影：齐欣

摄影：齐欣

摄影：齐欣

摄影：齐欣

西溪会馆 杭州
Xixi Club, Hangzhou
2008

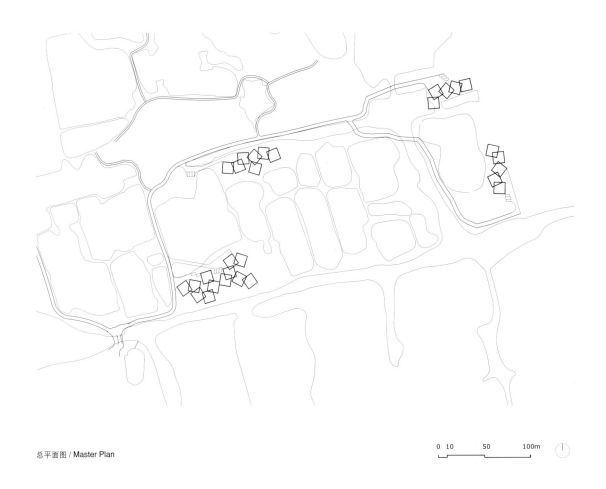

总平面图 / Master Plan

杭州的西溪被列为国家湿地公园，其中的两期已向公众开放，有着开阔的半人工化湿地植被和一些仿古建筑。在三期里，策划者想引进面向未来的建筑，招来十二名国产、半国产建筑师参与设计，内容无非是一些类似宾馆的休闲设施。

据说早在明、清时代，西溪湿地便被文人墨客们相中，有一搭没一搭地在此愤世嫉俗，吟诗作画。苏东坡、唐伯虎、郁达夫、徐志摩等前赴后继，留下早已消失的足迹。当建筑师们考察现场时，看到的是散布的鱼塘和农舍。农庄的形态在统一中蕴含着变异，或不土不洋，或既土又洋，散发出清爽的庸俗。

题目似乎清晰了：建筑与自然（可能更强调自然，淡化建筑）；建筑与历史（可能更偏重历史上的历史）。然而，建筑是否隶属自然？再者，何为历史？

人类启动了回归自然的征程，与祖祖辈辈愚公移山的事业大相径庭。但人在自然界里盖房子的本意是抵御自然，并非亲近自然。无论在东西方，自然界中的建筑历来都旗帜鲜明地标榜几何，与自然对抗。更何况，会所之类的建筑需要私密。因此，面对自然，采取"不回避但把控"的策略要远比"回避或妥协"更自然。

历史是一条延续的长河，她同属于昨天、今天和明天。当我们排除了厚古薄今的年龄歧视，客观审视文人墨客和当代村民的"遗产"时，发现二者间的共性在于：平面是简单的矩形，体形上切出了斜坡做屋顶，但简单的单元在组合中发酵，孕育出变化。对暂时幸免于难之农舍（短短一个月后，新农舍集体沦丧，倒是有一两间老房子得以幸免）的抽样调查显示，建筑进深在10–13米间徘徊。

于是，我们便踏着前辈们的足迹，将建筑类型定义成最简单的矩形平面：面宽＝进深＝12米，两坡顶，自由组合。为了跟简单较劲，甚至还锁定了"一个基本平面＋一个基本立面"的目标。

当12米×12米的单元平面相互纠缠到炽热状态时，咬合的部位生成出天井，它同时承担着采光与通风的使命，从而解放了外墙开窗的义务，或只在想开的地方开窗。

走到这一步，建筑开始自信，从容而轻松地坐在水塘边，树荫下，开始与自然推心置腹，促膝谈心：你是自然，我就不是么？我是建筑，你就不是么？这时自然忽然有所领悟：发现镜面瓷砖以破镜重圆的方式为建筑蒙上了一张魔幻般的外衣，它将天空、树木、水面乃至可恶的游人纳入本体，打碎，重组，然后再将升级版的客体影像释放出来，回归自然。

随着朝夕轮转、秋去冬来，建筑与自然已磨合到了难舍难分、交相辉映的境界，同呼吸，共甘苦。

功能	会馆
规模	5500m²
阶段	竣工
甲方	杭州市余杭区政府
乙方	齐欣建筑 + 北京筑都方圆设计公司
团队	齐欣 + 徐丹 + 王斌 + 张亚娟 + 罗彬 + 刘阳

Function	Clubhouse
Scale	5500m²
Phase	Bulit
Client	Yuhang District Government, Hangzhou
Design	Qi Xin Archi. + BZDFY
Team	Qi Xin + Xu Dan + Wang Bin + Zhang Yajuan + Luo Bin + Liu Yang

The project site is a national park with water as a dominant natural feature. The club will inhabit two or three families during the weekends.

The main idea is to conserve the previous buildings typology on this site as a memory. That typology, whether traditional or modern, is composed of a simple rectangular plan, covered by a sloped roof. The simple buildings units, once combined to each other, results in a varied and composite architectural forms.

The ceramic tile, as cladding materials, reflects the surroundings and transforms it into an abstract painting.

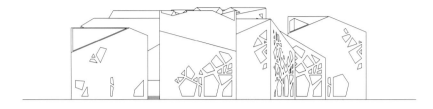

立面图 / Elevation

立面图 / Elevation

平面图 / Plan

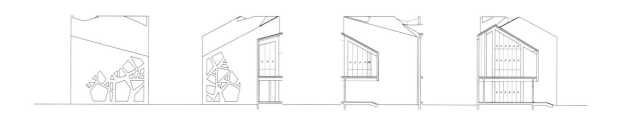

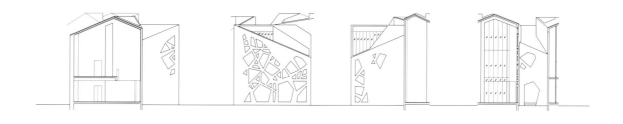

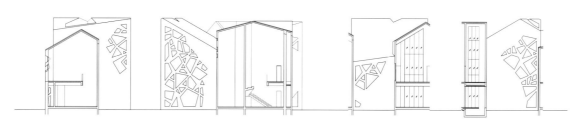

剖面图 / Sections

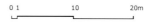

摄影 姚力

摄影 齐欣

撮影／青木

摄影：齐欣

摄影：齐欣

摄影：齐欣

摄影：齐欣

摄影：齐欣

摄影：姚力

Y-1-28 办公大厦 天津
Y-1-28 Office Tower, Tianjin
2008

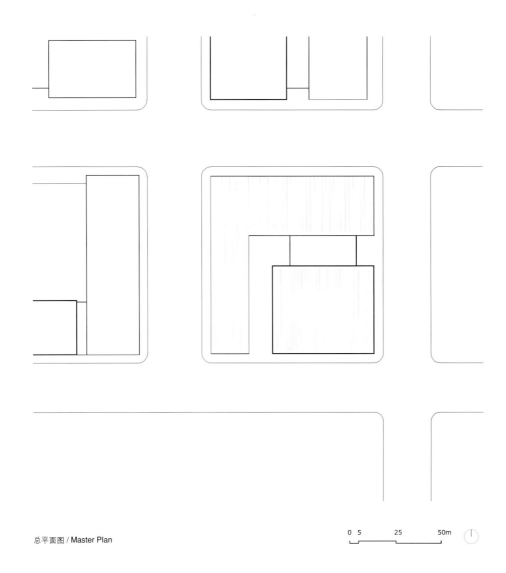

总平面图 / Master Plan

这还是一次集群设计,但其规模之大,前所未有:在一片新城金融区里,依照一张城市设计蓝图,九个建筑师各自设计一栋摩天楼以及它的裙楼。

设计导则详尽而严谨,组团的建筑风格被统一定位在"经典现代"和"出新不出奇"。

建筑的体量关系已定,无需更多思考。为了使组团内的建筑具有较好的协调,所有建筑师都在立面形式上做了同样的练习:横线条、竖线条和墙体上开窗洞。

本设计墙体上开洞的立面曾试图与结构相结合,以免除柱子对室内空间的干扰。立面的简单图案还试着以窗户出出进进的方式将南侧的绿地以抽象的形式引至建筑,但均匀的窗洞和窗间墙还是限制了隔断墙位置,使内部空间的使用不够灵活。

横线条和竖线条的开窗方式貌似解决了隔断墙的灵活布置问题,但事实上,由于立面与结构彻底分开,柱子会对内部空间的布局产生不利影响。

最终方案的结构选型由斜撑与竖向支撑相结合,立面的窗洞躲开结构构件,可将层高一分为二或一分为三。这一做法同时满足了消隐结构和室内灵活布局的要求。一分为三的小窗被放在西北方向,一分为二的大窗朝着东南方向。

本地块的西、北两个面都沿着商业步行街。因此,尽管处在东南角的主楼需要展示金融企业的庄重、大方,西北角的商业功能却要求建筑的平面和立面适当活跃,结合行人尺度,营造轻松宜人的环境。

裙楼的首层平面引入了内部街巷,行人可以自然而自由地进入到任何一个商店或主楼中。造型上,前一个方案试图将呈L形的裙楼化整为零,七个小体量的单体与完整而高大的塔楼形成鲜明的对比。后一个方案则在整中求变,将立面中的方窗洞分出五种不同的规格,其排列自然有机,犹如一幅画卷。画卷中的一些窗子伸出墙体,为平面的构图导入了立体的维度。

功能	办公
规模	122000m²
阶段	施工
甲方	天津新金融投资有限公司
乙方	齐欣建筑 + 天津市建筑设计研究院
团队	刘阳 + 齐欣 + 徐丹 + 王斌 + 高银坤 + 于向东

Function	Office
Scale	122000m²
Phase	Under Construction
Client	Tianjin New Finance Investment
Design	Qi Xin Archi. + TADI
Team	Liu Yang + Qi Xin + Xu Dan + Wang Bin + Gao Yinkun + Yu Xiangdong

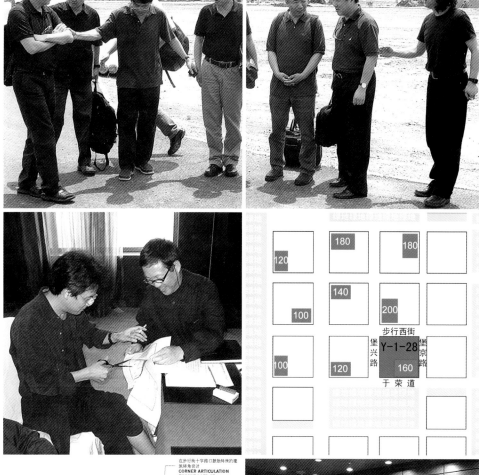

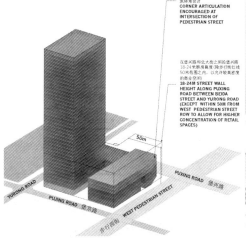

This is a rational tower building design, efficient and economical in cost. For efficiency, it combines the structure and the cladding system into one piece object, and therefore provides a working area cleared of columns, making the inner space more flexible. The square windows are arranged in a diagonal direction that allows the structural members to hide, and offers an optimal proportion of external wall between the solid and the glass part for energy saving. In addition, the windows at high level may drive the natural light into a very deep indoor space.

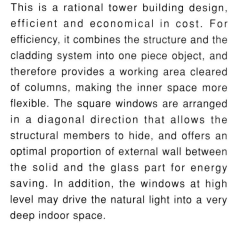

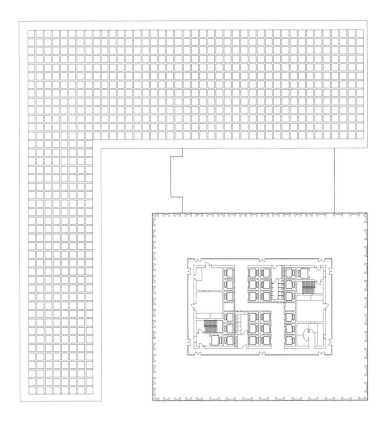

平面图 / Plan

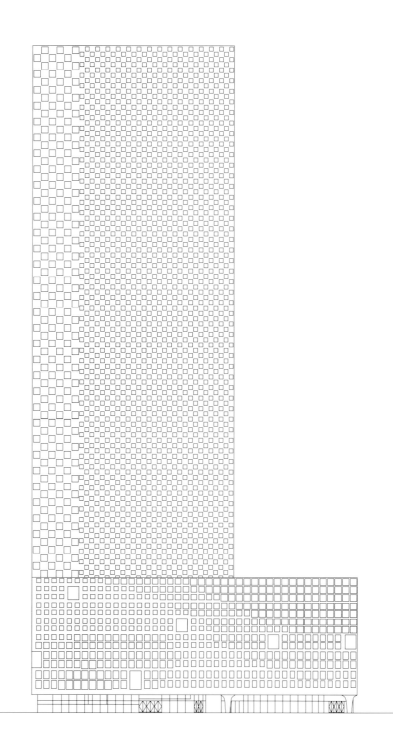

立面图 / Elevation

演艺中心　天津
Performance Centre, Tianjin
2009

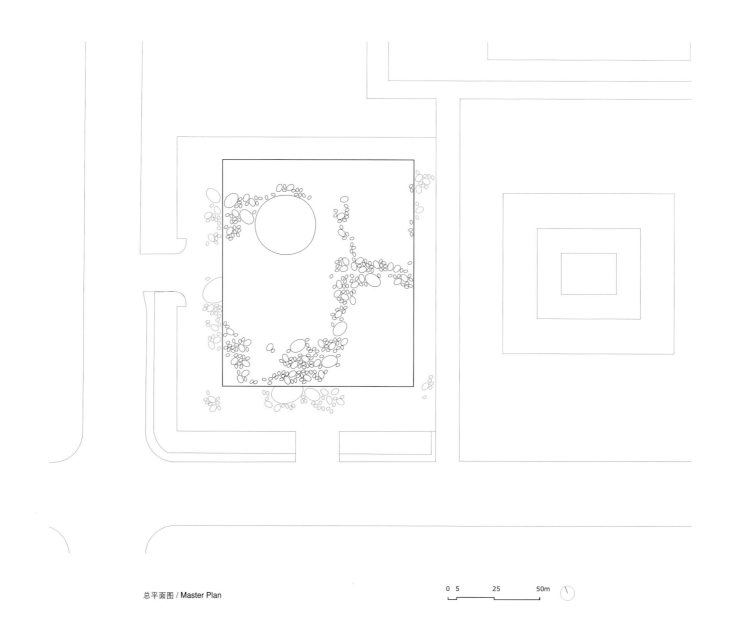

总平面图 / Master Plan

天津武清区的中心有一片比天安门广场略大一圈的"广场"，广场的北部是区政府，南侧是青少年活动中心，东西两侧排列着一系列政府职能办公楼。管它叫广场也不完全对，因为它的北部基本被绿化覆盖；中央将有一尊18米高的巨型雕塑，呈圆形；圆形雕塑再往南，是文化广场；这个广场的东西两侧准备各建一栋文化建筑，分别为文化中心（含图书、展览等）和演艺中心（含一个1200座的剧场和8个大大小小的电影厅）。

建筑师周恺负责设计文化中心，在他的推荐下，我们承接了演艺中心的设计工作。在和周恺的讨论中，我们调整了广场和建筑之间的关系，确定了建筑的尺度，并认为两栋建筑在共同框住广场的同时，形式可有差异，但应向广场一侧开放，形成广场的延续。

演艺中心坐落在文化广场的西侧。

剧场本身具有一套基本固化的流程，使其庞大而集中。为了强化演艺中心与文化广场之间的联系，方案将剧场摆在靠近城市街道的西侧，而将东侧的底层架空，让广场向这里充分渗透。架空的场所中不仅有从上空泻下的天光，而且为市民提供了可以遮阳避雨和便于夜间照明的户外活动空间。架在这一部位上方的是若干大大小小的电影厅。为进一步强化建筑与广场之间的互动，电影厅的坐席全部面向广场，使其在屏幕没有落下之前，可兼做广场活动的观众席。

白色是建筑外观的基本色调，中性的色彩易于在夜间利用灯光对它进行二次塑造。而一旦穿越这片白幕，半室外的市民活动场所以及建筑的中央大厅均以红黄两色作为基调渲染着演艺建筑的戏剧性。而红黄两色恰恰又是中国国旗的色彩，中国的色彩。

功能	剧场 + 影院
规模	42000m²
阶段	施工图
甲方	天津市武清区政府
乙方	齐欣建筑 + 北京市建筑设计研究院
团队	齐欣 + 戴伯军 + 刘阳 + 吴附儒 + 于向东

Function	Theatre + Cinema
Scale	42000m²
Phase	Working Drawings
Client	Wuqin District Government, Tianjin
Design	Qi Xin Archi. +BIAD
Team	Qi Xin + Dai Bojun + Liu Yang + Wu Furu + Yu Xiangdong

The Performance Center includes a 1200-seat theater and eight cinemas in different sizes.

The idea is to create interactions between this center and the public square on its side. The people can come through the building, underneath the cinemas. This open space provides an additional outdoor area where the people are prevented from the sunshine and the rain, and yet offers them the light at night. The movie halls are erected from this open-space, and their lounge faces the public square, just like the theatre's balcony, where the people can enjoy activities on the square.

立面图 / Elevation

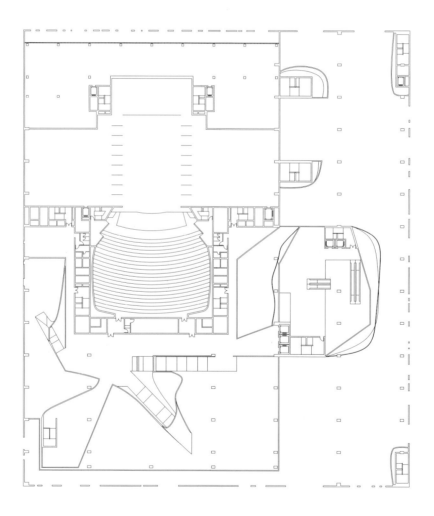

平面图 / Plan

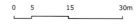

立面图 / Elevation

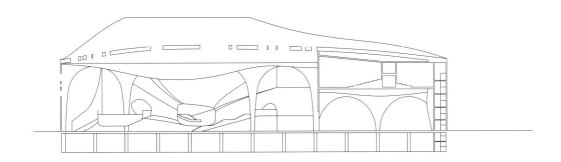

剖面图 / Section

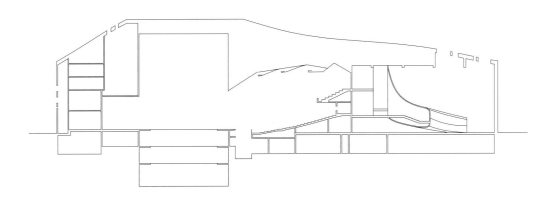

剖面图 / Section

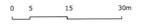

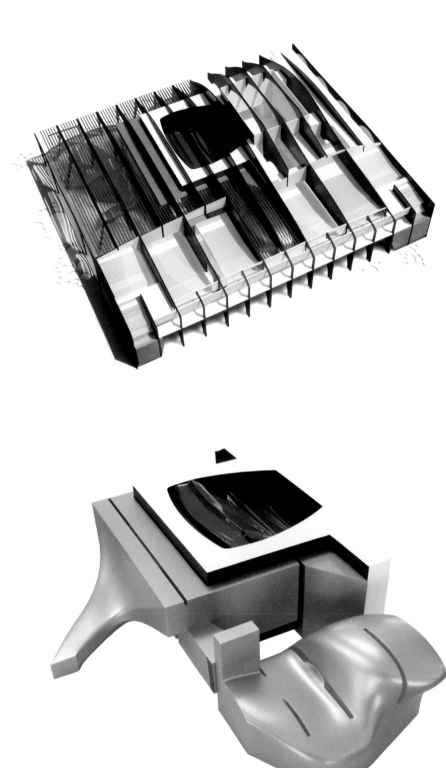

文化中心 塔什干
Culture Centre, Tashkent
2010

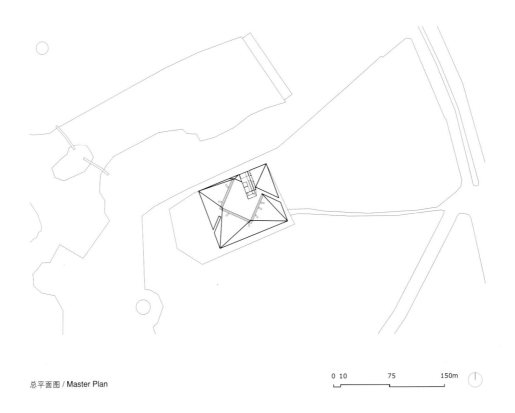

总平面图 / Master Plan

乌兹别克斯坦是个伊斯兰国度，曾隶属前苏联，其首都为塔什干。

项目紧挨着塔什干市中心的一个公园，僻静得似乎已经被人遗忘。地块里一米多高的杂草丛生，三座只有结构没有外墙的多层建筑显然已荒芜了多年。地块的西、南两侧被绿树、草坪环绕，东侧有个并不喧嚣的游乐园，北侧稍远处，隐约呈现出一片湖泊。

出于节省的考虑，业主决定将三个烂尾楼中靠南的一座保留，并将其作适当的扩展，创建一个文化中心。

进入这块场地，要通过南侧一块开阔的空地。从场地自身的位置看，来者先要接触到的是这个现存建筑的背立面，而正立面则面向北侧的一大片空地。于是，如何处理好这个既背（相对自身功能而言）又正（相对城市而言）的立面，成为我们所要思考的第一个问题。

现状建筑的体形是一个简单到没法再简单的矩形，从各个角度去看均相差无几，使这一没有特色的建筑变得可看和耐看，是我们要思索的另一个问题。

既然是文化建筑，就要在这一国度的传统文化中索取精神食粮。

方案先在场地中放入了一个清真寺中常用的方院。由于院子相对现状建筑转了个45°角，所以控制住了观望这栋建筑的不同角度。院子四周的房子只有一层高，屋顶被覆土，原来属于户外的活动场所并未被新建筑侵占。

像所有清真寺的院子一样，这里也有柱廊。为了更彻底地解决冬季寒冷、夏季炎热的气候问题，柱廊上悬浮了一面可升降的幕，可按照管理者的意志开闭。

处在院子中央的现有建筑的一角成为了视觉的焦点，这里的观光电梯不仅活跃了院子的气氛，还使访客能逐渐发现稍远的湖泊。

建筑的背立面如同伊斯兰妇女们所蒙的面纱。从城市一侧走来，看到的是一片朦胧，朦胧中带着诱惑，诱惑导引人们穿过夹道，进入楼体；或穿越楼体，发现院落；再穿出院落，走进自然。

功能	展览 + 餐饮
规模	2600m²
阶段	方案
甲方	匿名
乙方	齐欣建筑
团队	齐欣 + 戴伯军 + 于向东

Function	Exhibition + Restaurant
Scale	2600m²
Phase	Schematic Design
Client	Anonymity
Design	Qi Xin Archi.
Team	Qi Xin + Dai Bojun + Yu Xiangdong

This project is the renovation of a building. The client wishes to keep an unfinished building structure in the edge of a park located in the center of Tashkent, capital of Uzbekistan, and to renovate it into a Cultural Center.

The interesting thing is that the building stands face to the open space, with its back facing the urban street. Therefore, back and front are equally important.

The inspiration of this scheme comes from the Islamic women's veil.

Coming from the city side (women's back head), a curious and sensuous façade attracts people to explore the building. Through a narrow path, people will pass first under the building, then discover the courtyard on the other side, which again comes from the local mosque typology.

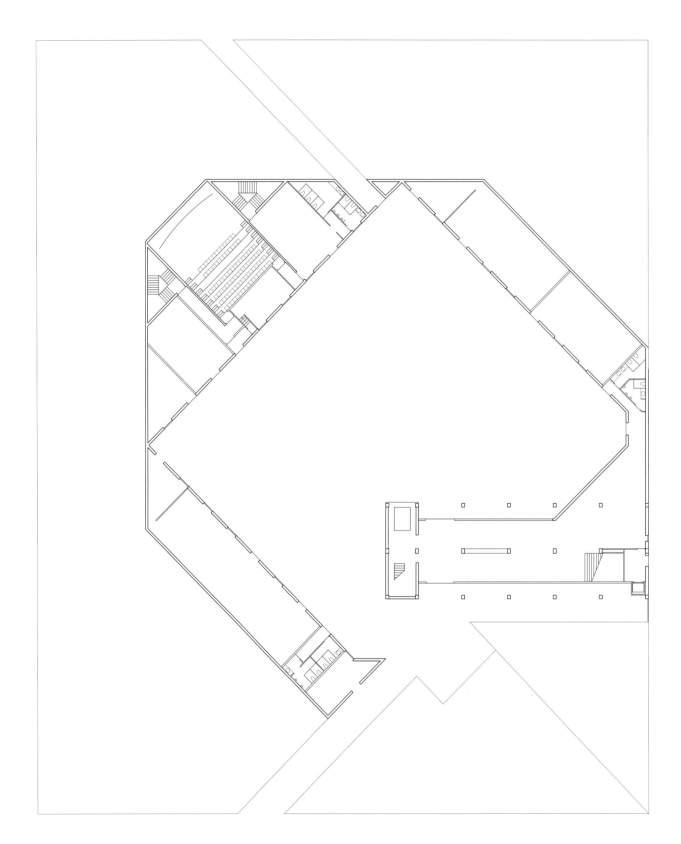

一层平面图 / The 1st floor plan

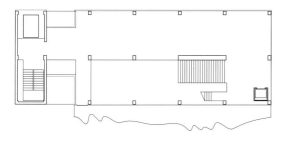

三层平面图 / The 3rd floor plan

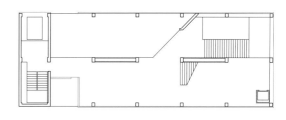

二层平面图 / The 2nd floor plan

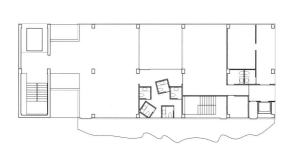

四层平面图 / The 4th floor plan

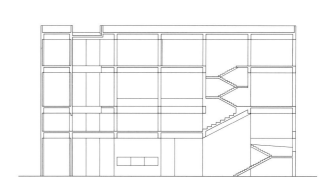

剖面图 / Section

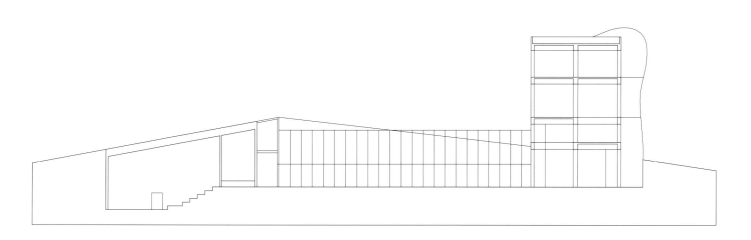

剖面图 / Section

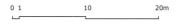

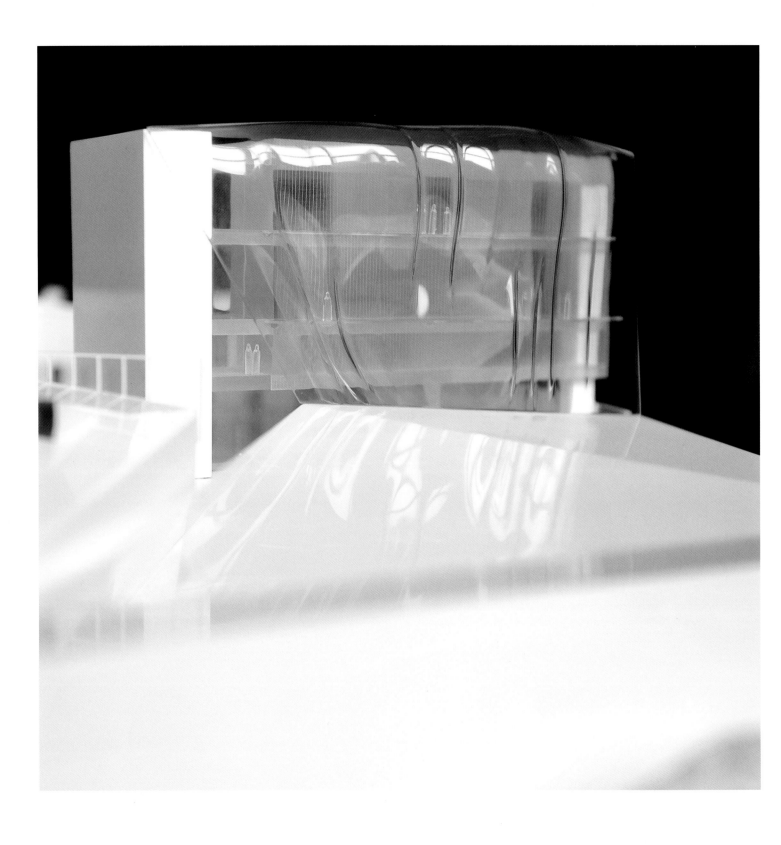

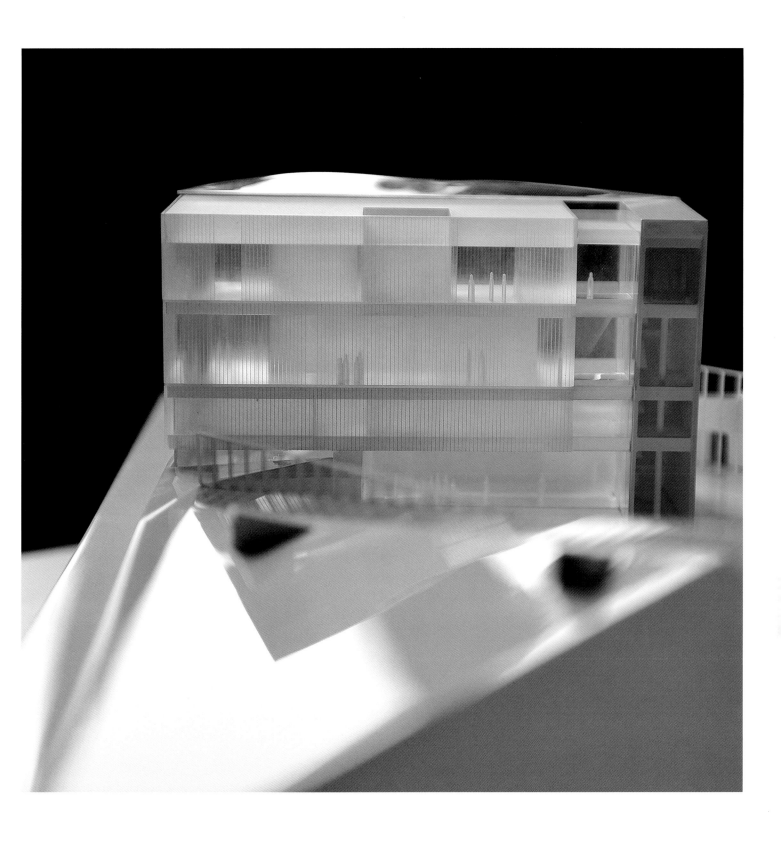

神农坛 株洲
Yan Di Temple, Zhuzhou
2010

总平面图 / Master Plan

都说曾经有个炎帝，那就算有了。有人说他老人家是湖南人，那就归株洲了。于是，株洲为颂扬炎帝文化，筹建着许多场所、场馆，其中便包括了神农坛。

神农，即炎帝也。传说炎帝有过许多功德：发现五谷，成为农耕的鼻祖；品尝百草，成为医药始祖；创立集市，奠定了贸易的基石；等等，等等。

神农坛，是祈拜神农的场所。拜者有心，不可白拜。于是有高人指点：分门别类，各取所需：想升学？升官？发财？丰收？还是长生不老？各有各自的包间。顺着这条线索，项目按阴阳五行分出金木水火土五个场所。每个场所，均以塑造意境为目标，回避用直白的材料语言来表达。

坛址被选定在一个已被毁掉了一小半的小山包上，旁边耸立着市里最高的构筑物——电视塔。山包的另一侧呈现着野长出的当代城市风貌，建筑高矮参差不齐。为了得到一片相对纯净的土地，并将山包恢复到原始的状态，祈拜场所们自觉地沉入了地下，或半地下，彼此之间由甬道相连。甬道分成了天、地、世三类：天道无顶，暴露出明净的一线天和山包上的树梢；地道黑暗，却也有一丝自然光洗亮了侧壁；世道有顶，但可平视旁边的草坡，与世隔而不绝，分享着自然界中的鸟语花香。不同明暗的甬道，作为引子，在铺垫着将至场域的氛围。

每个祈拜场所的平面都是方形的，其中四个小祈拜坛甚至是方体。方，反映着单纯，也标识着人类与自然的差异。往小山包上爬着爬着，一条小岔道处出现了一块掀起的方石板。从这里，便进入了一串连环的祭拜场所。

最先进入的是一个浴场。穿越一缕缕光线，尘世的心灵得到净化。随后，顺着阴阳五行的次生关系，各类祈坛循序展开。木，以草药为提示，颂扬着神农的医学成就，祈盼人们的安康。火，昭示着力量，黑暗中徒然映射出由自然光塑造的火红太阳。土，是神农之本。场所不仅在尺寸上加倍，更在视觉中增至无穷。无穷的疆土上，生长着养育众生的庄稼。金，是富足的象征，也是贸易的动力。闪烁着耀眼光芒的金属片，犹如灿烂的星空，或从空中徐徐飘下的落叶。水，在东方哲学中象征着智慧，它用无形促成变化莫测的有形世界。

场所是独立的，可根据选择出入。鲜花取代了香火，让五彩芬芳的花瓣洒落在祈拜的坛中。坛建完了，山包也随之恢复。在山包上，游人依然可以根据以往的线由散步、观光，与祈拜的人流互不干扰，和平共处。

功能	祭拜
阶段	概念
甲方	株洲市政府
乙方	齐欣建筑
团队	齐欣 + 戴伯军

Function	Worship
Phase	Concept Design
Client	Zhuzhou Municipality
Design	Qi Xin Archi.
Team	Qi Xin + Dai Bojun

In the Chinese legend, Emperor Yan (Yan Di) discovered the first grains, and is known to create agriculture; by tasting various herbs, he created medicine; he initiated market to create trade, and so forth. Together with Emperor Huang (Huang Di), another important figure of Chinese mythology, they become two ancestors of the Chinese people. Believe it or not Zhuzhou city is supposed to be Emperor Yan's home, which is why they want to celebrate his glory, and make it a place of worship.

The scheme follows the five elements in the ancient Chinese philosophy, creating five distinct places, each representing a particular Yan Di's merits and virtues: wood for health, fire for power, earth for good crop, metal for richness, and water for intelligence. None of them are interpreted directly with their material appearance, but with their spiritual meaning.

The places are always underground, sometimes open, sometimes closed, to keep the nature topography and landscape untouched.

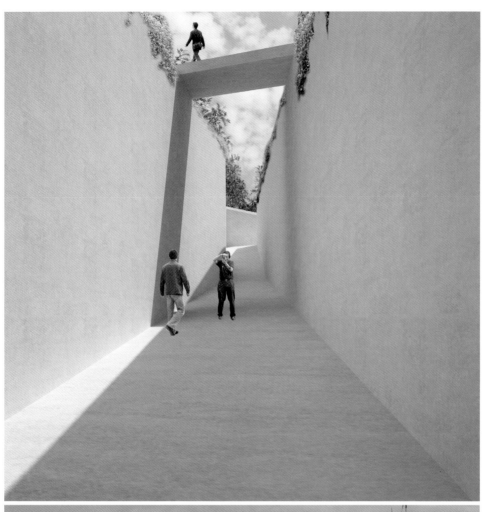

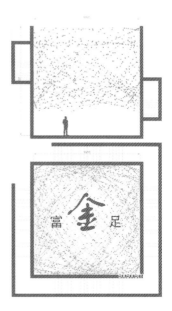
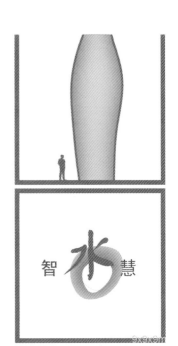

20+10 酒店 鄂尔多斯
20+10 Hotel, Ordos
2011

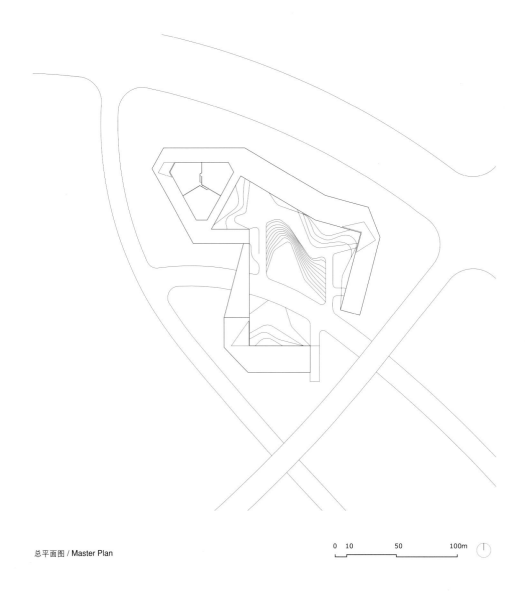

总平面图 / Master Plan

　　顺着一条河道，鄂尔多斯20+10项目在一片延绵的丘陵地上铺开。项目的主体是小办公楼。2010年，先由30位国内外的建筑师设计了60种形态各异，却又相互关联和默契的办公建筑。随后，又有新人加入，加入到被两条办公群带挟持着的公共区域中，营造与办公配套的各类服务性建筑。

　　由于人多，易乱，所以20+10的办公建筑设计受到了严格的限制。例如：房子一定要在一个特定的方框内做文章；在河滩地上的个子要矮，在沿城市干道一侧坡地上的逐渐长高。至于公建，城市设计却在鼓励它们的差异性，以活跃整个系统。

　　这条狭长的地块，从东南到西北，一共长了六个小包，其中的五个山包已被办公占领。由于一条飞架着的城市道路可能将老五和老六分开，最西北头的那个山包被空了出来，成为20+10酒店的落脚点。这酒店，其实还要兼会议、餐饮、娱乐和购物等其他办公区域的配套功能。

　　既然是20+10项目的一部分，就要与批量的办公建筑相关；既然不是办公建筑，就要与办公建筑有所区别；既然地处20+10项目的尽端，就要有开始或结束的意思。

　　酒店的总图像是一块马蹄铁，扣住了整个区域的端头。马蹄铁的开口面向20+10办公区的中心绿化带，高架道路可能对其产生意向上的分离，而在视野和行动上，却紧密相连。

　　延续的建筑犹如哈达，飘落在场地的上方，盘旋中，缔造出大小不同的庭院。尽管内向的围合处理与散点式的办公布局形成对比，建筑本身的高矮却完全依循了办公建筑"高地上高、低地上矮"的设计原则。

　　贯穿20+10场地中央的道路，在这里将酒店与商业两大功能区自然分开。被围在建筑之中的山包，以其常见的姿态呈现出浮云与梯田。浮云下面，是几个主要功能区的入口；梯田，构成了自然与人工景物间的媒介。

　　以内蒙当地沙土制作的外墙板，带着微妙的色差，为建筑穿上了绿色的外衣。哪怕在凛冽的寒冬，这里仍荡漾着徐徐的春意。

功能	酒店 + 会议 + 娱乐 + 商业
规模	54000m²
阶段	方案
甲方	鄂尔多斯东胜区区政府
乙方	齐欣建筑
团队	齐欣 + 刘阳 + 于向东

Function	Hotel + Confrence + Entertainment + Retail
Scale	54000m²
Phase	Schematic Design
Client	Ordos Dongsheng District
Design	Qi Xin Archi.
Team	Qi Xin + Liu Yang + Yu Xiangdong

A continuous long building construction cuts off the noise from the urban express road. This linear wall is built up and down following the natural site. It embraces an open and peaceful area. Here, the canopies in forms of cloud float on the terraces, which are covered of colorful plants. The façade of the building are made of sand, a very local material, with a delicate tint of green.

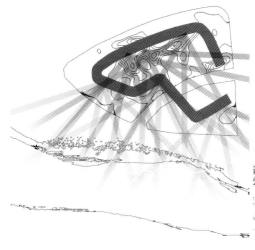

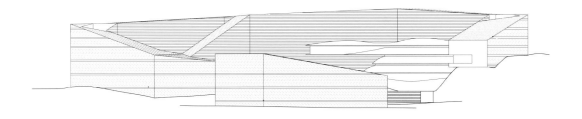

立面图 / Elevation

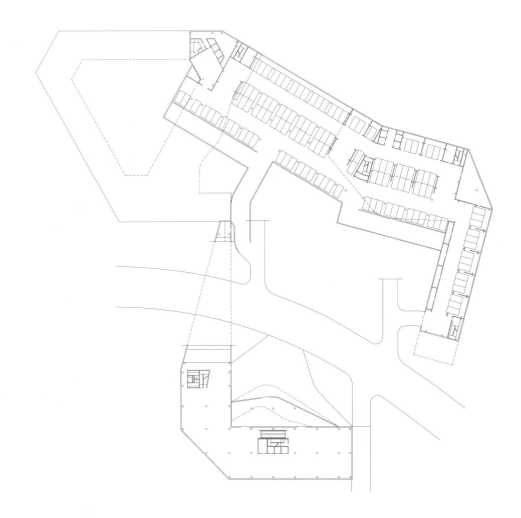

平面图 / Plan

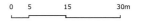

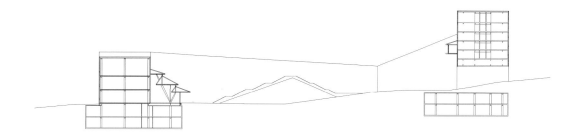

剖面图 / Section

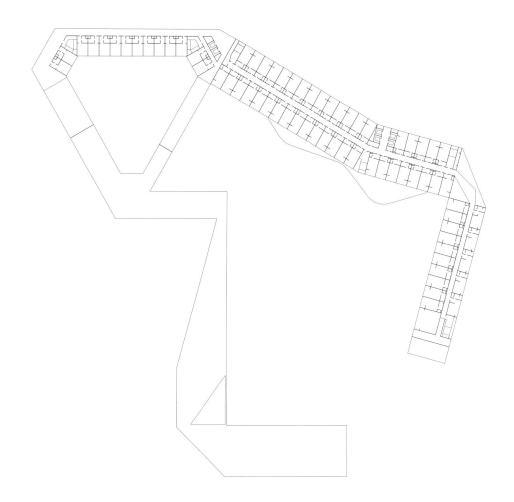

平面图 / Plan

132

135

学生活动中心 天津
Students Centre, Tianjin
2011

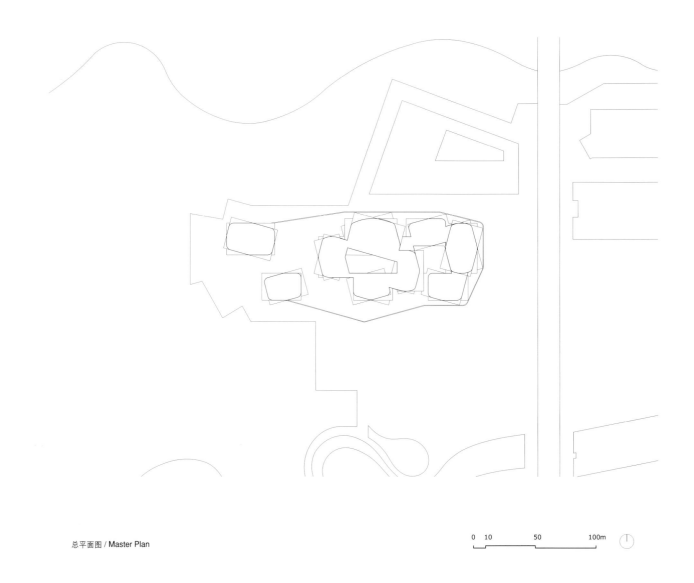

总平面图 / Master Plan

学生活动中心坐落在新校园主轴线的尽端，突出水面的半岛将其引向开阔的自然景观，与轴线序列空间中端庄的校园景象形成对比。

按照园中园的布局方式，中心由一组功能各异的小型3层单元体构成，并被一圈飘在空中的半透明的院墙拢住。院墙在尺度上与广场南侧的音乐厅和东侧的图书馆相呼应，却又并不阻碍内部与四方公共空间的渗透和相望。单元建筑彼此时分时合，各自的楼板在转动中形成雨篷或阳台，暗示着中国传统建筑中屋檐的出挑。靠近水面的建筑挣脱了围墙的束缚，跨向湖区。被挣断的院墙无奈地在广场一侧打了个折，从而为中心的入口划进了更为从容的前场。园中园的地面延续着校园建筑的红砖，而屋顶，却满铺了太阳能集热板，自力更生地解决了自身的能源需求。

整组建筑通透、飘逸，在分与离之间构筑各种供学生交流用的平台、连廊和院落，吻合了"团结、紧张、严肃、活泼"的八字方针。

功能	活动 + 办公
规模	10000m²
阶段	方案
甲方	天津大学
乙方	齐欣建筑
团队	齐欣 + 刘阳 + 于向东

Function	Activities + Office
Scale	10000m²
Phase	Schematic Design
Client	Tianjin University
Design	Qi Xin Archi.
Team	Qi Xin + Liu Yang + Yu Xiangdong

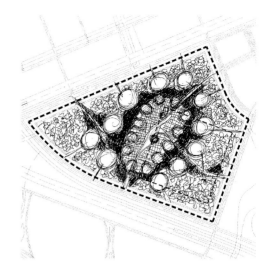

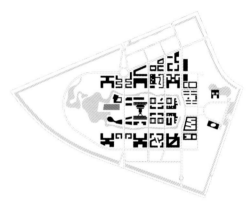

The traditional Chinese garden art consists in putting the simple unit of buildings into a vague or narrow space, at random, to result in a picturesque vision.

When looking at the building concept in Europe, the building reduces volume as it is built up. But in China, the building is constructed with the ground plan often smaller than the roof, which cantilevers beyond.

The Students Center is treated as a Chinese garden in the campus. Each level is built up and twists, forming cantilevered floors, which are oriented differently. As the building twists on its core, the floor slabs may touch each other with their neighbor's slabs at the same level. The corners of every flat may then be a balcony, a corridor, the shading louvers or the entrances canopy.

A suspended translucent belt in aluminum encloses the buildings as a fence, just as in a Chinese garden. This symbolic enclosure allows peoples to access freely to the building, and harmonizes with the scale of the big concert hall beside.

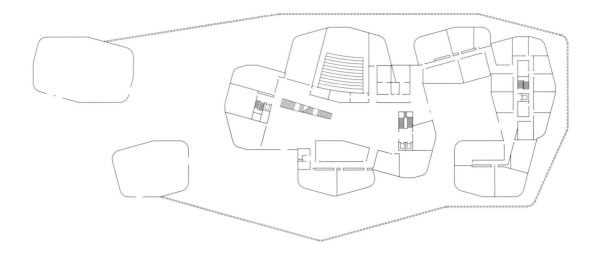

一层平面图 / The 1st floor plan

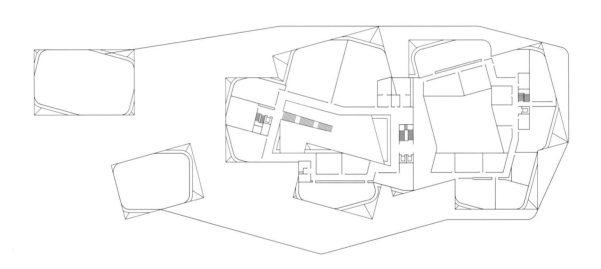

二层平面图 / The 2nd floor plan

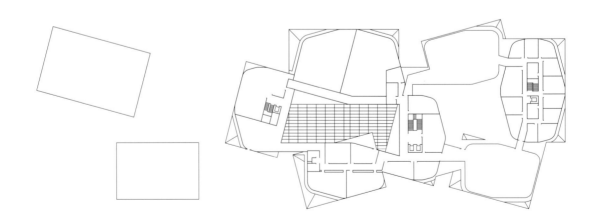

三层平面图 / The 3rd floor plan

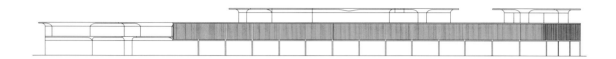

立面图 / Elevation

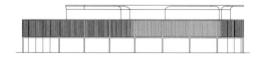

立面图 / Elevation

剖面图 / Section

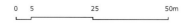

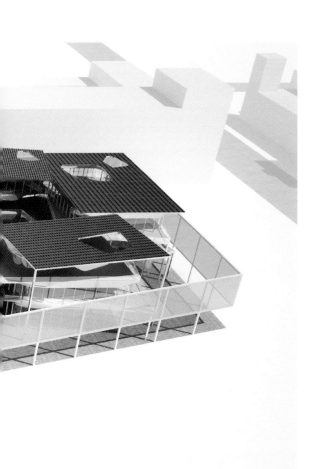

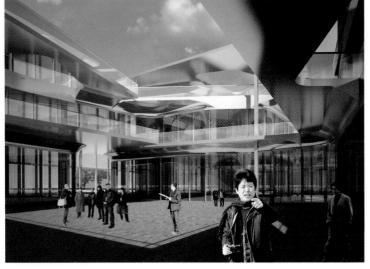

地铁网控中心　　昆明
Metro-Traffic Control Center, Kunming
2011

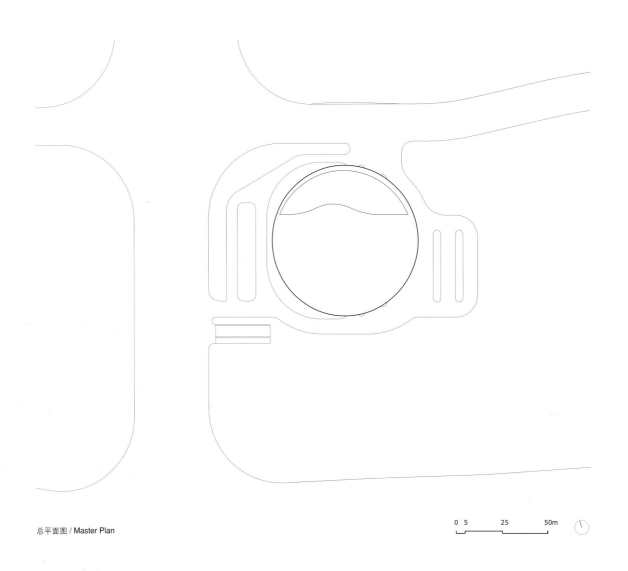

总平面图 / Master Plan

与通常城市里的大型公共建筑一样，昆明市对这一建筑的外形具有极高的期盼。但这又是一个有别于通常办公建筑的项目，因为它的功能性极强。所谓功能，除了指那些由工艺产生的繁杂技术要求外，更具挑战性的是如何统筹组织几股不同而又大量的人流。

第一股人流来自地铁。不久的将来，将有三条地铁线在此设站。

第二股人流来自公交。地块原本就是一个公交总站，未来仍将有 12 条公交线在此停靠。与其相关的是公交的调度、司机的休息等。

第三股人流来自地铁还建给公交的办公人群。

第四股人流来自地铁线网控制中心。它包括自身的办公人群（其中不同办公区域有不同的安保级别。最核心的区域是它的中央控制室，需要开阔的视野和高大的空间），并且需要经常接待来宾、要客。

与上述人流相关联的当然还有机动车与非机动车的停放，出租车的停靠和各类旅客的换乘。

最后一股潜在人流是不受欢迎的，那就是来自恐怖主义的威胁。因此，要尽可能通过建筑设计的手段规避各类危险，以确保城市交通系统的安全运营。

难点集中在地块本身小得可怜，且周边的路网也相当复杂。

通过若干轮方案的比选，我们最后推荐了这一不仅在垂直和水平两个维度综合解决了交通的问题，而且用简单明了的形体将各类功能统辖于一身的方案。

方案的顶部放下了要有高大空间的中央控制室。建筑主体的内部被分成还建和自用两大部分。

尽管两者的核心筒背靠着背，但其余部分被通高的中庭分开。中庭同时解决了各自的采光与通风需求。

地面层，在迎向城市的一面，设置了两个办公楼的各自入口。入口背面潜伏着大面积的公交站，无形中将建筑托举起来。

地铁与公交的换乘、机动与非机动车的停放均被井然有序地安放在地下层中。

建筑的体形是一个简单的圆筒。立面的图案令人联想起云南少数民族的传统编织物、谷仓、服饰等。它将现实拉回到原始，又将原始引向未来。

功能	公交枢纽 + 办公
规模	69000m²
阶段	扩初
甲方	昆明轨道交通有限公司
乙方	齐欣建筑 + 铁四院
团队	戴伯军 + 齐欣 + 于向东

Function	Trasportation Centre + Office
Scale	69000m²
Phase	Preliminary Design
Client	Kunming Rail Transit Co.Ltd
Design	Qi Xin Archi +CRCC
Team	Dai Bojun + Qi Xin + Yu Xiangdong

Beyond the difficulties of the technical and security requirements for this building, the real challenge comes from the traffic circulation. In the underground, there are three subway stations; above the ground, there is a twelve-line bus terminal. The office building needs to be divided into two parts: one is for the subway network control center and one for the bus company, with their independent entrance and car parking.

Although the plot is very small and tight, the pedestrian must be separated form all the different vehicles (taxi, bus, fire engines and bicycles) for security reasons. The final architectural form is just a pretty pure cylinder, slightly raised from the ground, with a crossing pattern that refers to the local traditional fabrics.

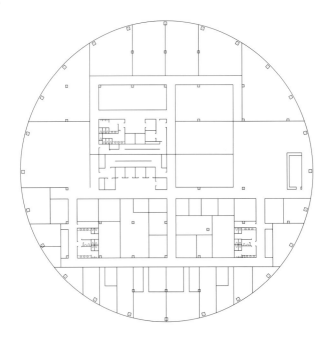

平面图 / Plan

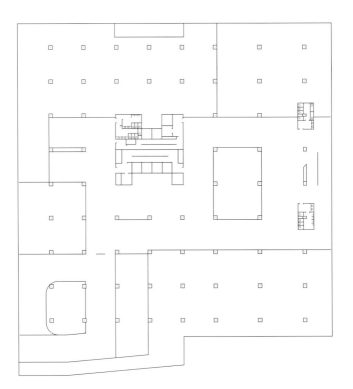

平面图 / Plan

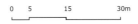

剖面图 / Section

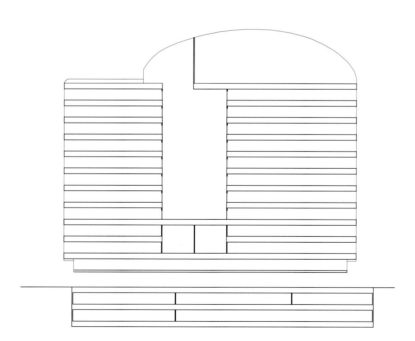

剖面图 / Section

齐欣访谈
Interview

采访人／黄元炤
北京　2011.9.30

黄：您在境外事务所的工作经验非常丰富。其中有法国巴黎建筑与城市规划设计院（SCAU），能谈谈这段经历么？

齐：我在SCAU工作时，那里有五个合伙人，实际上是五个小事务所共享资源，我只跟着其中的一个叫Zublena的合伙人干。事务所里名气最大的是另一位合伙人，叫Macary，他很会做关系，所以境外来的大腕都跟他合作，有丹下健三、贝聿铭、库哈斯、博塔等。这样，我就间接认识和接触了一些当时的知名建筑师。

我的老板也教书，并成了我做法国建筑师文凭毕业设计的导师。由于他整天忙于经营和教学，所以设计的事更多由下面的人完成。我先是以临时工的身份进去的。清华的训练使我能解决一些实际问题，所以一个月后就转正了。那时，老板安排我跟一个叫Tomas Sheehan的美国人搭班。他早我两三个月到，管创意和造型，我来落实平面以及和各工种的配合。我们俩一起投过两个标：一个是斯特拉斯堡科技园，一个是南特矿业学院，都中了，也盖起来了。这段时间里，我天天画平面，练了基本功。那个美国搭档属于特飘逸的那种人，瞎画一根线就特漂亮。他那松弛的态度和艺术的风范对我日后的工作产生了一定的影响。对受过工科教育的人来说，严谨容易，飘逸难。要跳出严谨的框框，其实是件挺有挑战的事。但有了工科的基础，又益于将飘逸的东西落到实处。

黄：自1994年起，您又在福斯特的香港公司工作过三年。福斯特属于有高技倾向的建筑师。据我观察，您早期的作品也有技术主义的倾向。比如北京的国家会计学院，您用虚化的玻璃与金属搭配，创造出精致的建筑和新的空间感观。东莞的松山湖管委会项目也有此类倾向。这是否受到了福斯特或其他有技术主义倾向的建筑师的影响？比如努维尔？

齐：把福斯特归类于高技派，也许因为他的成名作是香港汇丰银行。这个建筑还让他背上了一个黑锅：好像他只会做贵的建筑。在福斯特那里，我发现他有几点特别自豪：一是不向业主说不（业主提出的要求是设计的起点，在此基础上出绝活儿才是高手）；二是不超工期；三是不超预算。有一次福斯特得意地给我们介绍一个在英国做的厂房，外墙用的是瓦棱板，形式简单甚至简陋。他得意，是因为业主的限价极低，谁都做不出来，他却做出来了。汇丰银行贵，是因为业主要求做个全世界最贵的建筑。当年，在《中英联合声明》签署后，香港资金外流得很厉害，而汇丰银行占有香港90%以上的资金。为了表明他们不打算走，以留住资金，安定人心，汇丰就要盖一个最贵的房子。此外，他们还要求做一个可迁移的房子，万一香港没戏了呢？这才引出了福斯特的高科技建筑。

从汇丰银行起，福斯特就同做工业设计的设计师合作，后来其中的一两个人还成了他的合伙人。"高技"实际有两层含义，一个是"高"，一个是"技"。"技"只是指将功能构件视作美的东西，未必"高"。按这一标准，从欧洲的哥特式建筑到中国的古建都属这一派。而"高"，则相对于"低"。福斯特并未一味追求高端技术，也不刻意夸张技术的表现力，而更注重于利用现有技术和对结构的忠实。对我来说，他是不是高技派并不重要，重要的是他把大小项目都当成一个物体去做。香港的赤腊角机场是当时事务所里最大的项目，也是当时全世界一次建成的航站楼里最大的一个，而那时事务所里所做的最小项目，是给一个意大利的五金商设计门把手。两个项目都是一笔做完，整体感非常强。

和法国不同，英国人做设计很讲求逻辑。出手前，要搜集很多资料，进行分析、推理。在中国，推理多了就容易僵化，做不动了。而福斯特的强处恰恰在于无论有多少限制，都能在遵循逻辑和满足限制的前提下，做出一个意想不到而且非常富有诗意的建筑。

跟像罗杰斯这样的高技派建筑师相比，福斯特的设计更强调整体感，更优雅。皮亚诺和福斯特有点像，不光整体感强，而且有诗意。哪怕到今天，哪怕有时我会先将物体剁碎，也不会忘记建筑的整体性。这多少来自福斯特的影响。

我喜欢有逻辑、有灵气、有诗意又大气的房子。

黄：确实，您做的北京国家会计学院让人感受到一种典雅、简洁、稳重和大气。

齐：在北京国家会计学院项目上，我受到了双重的影响：前区公共建筑的影响来自福斯特，后区的宿舍建筑更多受到的是法国的影响。

黄：1997年，您参加了北京SOHO现代城的设计竞赛，当时参赛的还有崔愷和朱小地。您在方案中提出了绿色、节能的概念，利用太阳能，并将绿植引进超高层建筑。之后，您也做了些生态城的规划和设计。有些规划项目还找景观团队合作。从中，我注意到您对自然、生态、节能、环保以及可持续发展的议题感兴趣。您能对这部分做一些说明吗？

齐：绿色建筑是人们普遍关注的一件事。我在这方面的关注跟福斯特有关，因为他是最早获得欧洲专项资金去研究小能耗与绿色建筑的建筑师，并开发了很多相关的技术。但设计毕竟是一件综合的事，设计师要关注方方面面的问题。对建筑师来讲，只要有这个节能、环保的概念就行了，剩下的事要请更专业的人来完成。把绿植引进建筑，也只是个概念。现代城在长安街的南侧，露在街面上的是北立面。这么大的墙面常年不见阳光是件很背的事。为了把正北立面缩到最小，我在北侧画一条弧线。这样，自然光就几乎能进到所有的办公室了。做总平面也能节能，这不需要什么技术。

知识分子应始终抱有批评或批判的态度。某些事，别人都不想，你就要去提醒；而当大家都在做的时候，你就不用太关心了。现在，是人都在讲绿色建筑，讲节能、环保、低碳。我觉得在这方面的使命已经完成了，可以去关心别的事了。

黄：如果说北京国家会计学院的设计有高技倾向，您后来的设计似乎就有了明显的转向。比如北京香山的传城和东莞松山湖的管委会。这两个项目好像更关注建筑与周边环境的结合。2007年后，您似乎又关注起建筑的表达，倾向表象性的设计。更严格地说，在2003年的和平丽景项目中，您就有塑造商业街区立面的不同表述，2004年的用友总部也思考到窗洞的有机组合。您怎么看这些转变呢？

齐：像所有国际大腕的事务所一样，在福斯特事务所里，他的设计几乎变成了一种宗教。身在其中的人都觉得只有这玩意儿好，其他都是狗屁。从那里出来后，要花很长时间去逃脱福斯特的阴影，这是自觉或不自觉的事。转变是自然、必要和漫长的。

黄：当时您似乎想很快地跳出来。

齐：至今我也不认为福斯特不好。但被他的阴影笼罩，就找不着自己。除非你坚持认为他就是上帝，永远按照他的方式去做，那你就永远是上帝的奴隶。福斯特在20世纪80、90年代是世界上非常重要的建筑师。到了21世纪，新出现的建筑明星是赫尔佐格和德梅隆，他们在打表皮的主意。你会问，自己是不是掉队了？现在，扎哈又把人们的目光引向造型。要不要跟着动呢？

人总受到方方面面影响。一方面这表明你还没找到自己，一方面证明你还活着，否则就按照一种方式做就完了，想都不用想。我从来不那么自信，也不想在一棵树上吊死。看到这些年建筑潮流的起伏，会感到任何事物都有好的一面，但又不绝对。跟风显然没意思，但旧的东西对我又没刺激。所以，每次都在测试新的途径。这种活法未必轻松，却有意思。

黄：算是喜新厌旧？

齐：从事创作行业，喜新厌旧是必要条件，否则就别干了。你要卖包子，就每天做同样包子去卖就完了。

黄：我认为您转向了对表象、表皮的操作与表述。比如北京的贝克特厂房，在一个立方体上开大小不一、随机排列的方窗，形成抽象的几何图案。这是对表皮的单纯思考。武汉融科天城的售楼处，您在建筑外围蒙了一层表皮，构成内部空间与外部环境之间的介质。这是对表象、表皮正反两方面的思考。江苏软件园项目中，您在一个方正体外加上了一层竹帘及花棱窗；天津鼓楼区的商业项目，您用细柱创造出虚实相生的立面。这都说明您从高科技的设计转向表象、表皮。南京风情街其实也是一个倾向于表象、表皮的设计。您自己如何看待这个转向？

齐：你说的不全错，也不全对。当下的中国很浮躁，我又属于自己不会找项目的那种人。人家找你，就期待你做出吸引眼球的东西。这时，所谓的表象就变得重要了。这一现象叫作逼良为娼。几年前，我的小孩在法国幼儿园里的一个小朋友的家长弄了块地，要盖房子。我利用春节长假，按功能要求给他们做了个很简约的设计：除了平面是个三合院外，该开窗的地方开窗，该开门的地方开门，立面简单朴实，结果就盖起来了。建筑其实做到这一步就够了。但你在中国要这样干，人家就说你没干，或没能耐。

具体问题还要具体分析。

我去武汉接洽天城售楼处项目时，开发商说这个房子是广告，一定要吸引眼球。还说做什么风格没关系，但千万别做中式的。在武汉，特别是汉口，都是些殖民地时期留下的折中主义房子，确实跟中国没啥关系。售楼处旁边还将起一堆超高层住宅。因此，只有把房子做整了才能大，大了才不至于被那些巨魔镇住。剩下的是如何把建筑做花，不花人家就不找你。当时景观设计的概念已经出来，铺地用的是冰裂纹图案。而把冰裂纹图案用到建筑的立面上，正好还能解决西晒的问题。开始，我想找一百个老太太用柳条去编个大筐，将编筐本身做成行为艺术。结果没弄成，倒让世博会的西班牙馆实现了。最终，冰裂纹的图案是用铸铝做的，因为武汉那边有造船厂，让他们做这东西很容易。

北京的贝克特厂房则恰恰相反。业主是美国人，起先我还担心他能否接受中国概念。但图一出来，他欣然接受。人家在美国待腻了，来中国就想换个口味。你光关注表象了，将厂房人性化、地方化、院落化才是这个方案的实质。此外，如何解决大跨厂房里的采光是我关注的重点。方案利用了金属桁架的斜撑，错落布置出能起到均匀采光作用的侧高窗。从屋顶上看，成了副棋盘。屋顶的方形图案逐渐散落到立面，并开始跳跃，打破了以往厂房建筑的呆板。方形的办公楼，方形的庭院，方形的窗子和方格的屋顶构成了一个整体。

黄：我还是认为您在表象表皮上做了不少的思考与尝试。在秦淮风情街中，为与当地发生关系，您把那里的民居聚落用抽象化的图案反映在墙面上；在杭州的西溪会馆中，您开始探讨表皮与空间的关系，形体挤压出的天井解决了采光通风的问题，被解放了的外墙用树枝状的图案来隐喻自然山水；而良渚玉鸟流苏的商业街，也是经由形体挤压创造出的一种形态，同时衍生出带状柱廊般的骑楼和街道。西溪会馆与玉鸟流苏似乎都在用整合碎片的手法。您怎么看这两个项目？

齐：你总盯着建筑的外表。秦淮风情街项目的灵魂不是那张皮，是规划。当时有个总规，街道是直直的，广场是集中而庞大的。我其实是在用房子挤压街道，制造建筑之间的张力和有意思的场景。这是个夜场，夜场有意思的地方是不断发生意外的故事。这里的街道忽窄忽宽，一会儿在地上，一会儿在空中，时而在建筑里，时而又转到建筑外。目的都是拨正反乱，把人弄晕。让人们在微醺的状态下摩肩擦背，巧结良缘，巧逢艳遇，簇生出拥挤的城市魅力。这是一种城市策略，是对任务的解答，而不是表皮这么简单的一件事。

玉鸟流苏是在做一条商业街。但我只负责街道的一侧，另一侧归张雷管。雷爷压根就不想做街道，或只想做他自己内部的街道。我的地块呈带状，非街不可。问题在于如何既顾及建筑的整体性，又反映不同商铺的个性。骑楼前的细柱在人视点上塑造了街道的整体感，而且上面可以随意悬挂广告。柱廊后面的房子在扭来扭去，其内部空间也各不相同，反映着每个商铺的个性。一条延续的屋脊，把扭动的房子们又串了起来，再一次建构了整体。这是城市的概念，有别于简单的单体建筑设计，或如你所说的表象、表皮设计。想着城市的事去解答具体问题，才能相对周全。

杭州西溪的会馆设计不需要跟别的建筑呼应，只要关注自然。我认为不应向自然方面去靠。因为再怎么靠，也做不过自然，而且是伪自然。所以，我干脆就坦荡做人，做人工，做几何。但几何形体被附上了光洁的表面，天空、绿植映到上面后被反射出来，被变形，被抽象化。随着季节或时辰的变化，建筑可以即时地与自然对话，有点像同声传译。人还在，自然也在。

黄：用镜面转换的手法很抽象，也很中国。从镜面中不仅可以看到自己，也能看到四周景物，而四周景物代表着现实，也代表着过去，好似把过去与现在融为一体，成为凝固的意象，就如同绵纸的迷蒙、铜镜的朦胧、玉的透与不透、书画中的淡与浓墨或线到点之间的层次。这种凝固体中产生的迷蒙反映出一种文化和这一文化的深度。这种由镜面表皮引发出的抽象意境是否是一种对中国性的表述？

齐：我没你想得那么多。如果这个项目中有什么中国韵味的话，也是从命题来的。确实，以前文人墨客们在这里盖过房子，留下了诗篇。为回应这段历史，西溪的一、二期建设中做了些复古建筑。我们参与的是西溪的第三期建设。我拒绝只想明清这段历史。第一次去看地时，我们看到的是成片的农民新村。杭州一带的农民房特有意思，用浅灰和深灰的瓷砖错落拼贴，构成一种发晕的图面。这里的人还喜欢用不锈钢做防盗网和护栏，特别是把房顶的天线兼避雷针做成一个闪闪发光的不锈钢球。

历史是一条长河，不应说只有明清那段历史有价值、要继承，农民新村就算了。我把历史给拍平了，让前后辈们平起平坐，具有相同的尊严与价值。接着，再寻找他们的共性，以便踏着先辈们的足迹前进。不管明清文人还是当代农民，都用一些简单的几何体塑造房子，都是坡屋顶。单元在组合中的变化和错落，使群体变得丰富了。方案用一个12米×12米的方形作为基本单元，它还沿用了祖宗们喜欢的坡屋顶。单元在组合中发酵，产生了变化、混乱或自然。我也只设计了一个立面，上面全是洞。将这一基本立面用在方体的不同方位上。不需要开窗时，就把洞给堵上。

几何语言可以属于明清，也可以属于中华人民共和国；可以属于过去，也可以属于未来；可以属于中国，也可以属于世界。我并没刻意想中国这件事。但要说镜子，我印象特深的是阿姆斯特丹的河水。在水里，建筑变形了，树变形了，灯光变形了，人也变形了。我在用实在的物体构筑虚幻的影像，稍微脱离一点过于写实的现状。

黄：让我们回归建筑学，您有没有自己独到的设计思想或信仰？

齐：没有。设计理念是媒体关心的事，所以每个事务所都要给自己编个理念。让我一看，全对。只是有的偏重社会、政治或文化，有的偏重城市、空间或几何，有的偏重建造、技术或材料，有的偏重光，有的偏重色，有的偏重可持续发展，有的偏重服务，有的偏重人。你要把所有这些理念加在一起，就成了我的理念。

黄：那您的事务所的个性体现在哪儿呢？

齐：有些建筑师，比如安藤、盖里、扎哈等，有很强的个性。但我不想有什么东西能限制住我，想去哪儿就去哪儿。刚才你一不小心把努维尔搁到福斯特那帮人里了。但实际上，努维尔是个总在变的建筑师，你想不到他下一步要干什么。我更倾向于这种做法。

黄：您在避免落入常态。但看您的作品，还是一眼就能认出这是您做的，您的效果图也很有特色。这也是一种风格。尽管您拒绝固定的风格，外人依然能感觉到您风格的存在，而且还很明显。

齐：趁你不注意，唐僧给你画了个圈：那是佛掌，那是你所处的社会、你的阅历和文化。你永远跳不出如来佛的掌心。

黄：这是一种没思想的思想？

齐：也许吧。

黄：不能开条自己的路么？

齐：我一直在走自己的路，每一步都是自己迈出的。只是我更愿意随缘，所谓见机行事。所以我并不知道以后会往哪儿走。让每一个项目的具体情况诱导出为其特制的、合宜的建筑。

黄：最后一个疑问——写历史时，人们会把某一时段的建筑分成若干流派，再让建筑师们对号入座。您不给自己订座，是否想让历史学家换一种方式来描述历史？

齐：恐怕很少有建筑师喜欢被人贴上标签。评论家和建筑师最好还是各行其是，各尽其职。

H: You have a lot of experience working for overseas design firms. One of which is the S. C.A.U in Paris. Could you tell us about it?

Q: At the S.C.A.U, the office was composed of five partners. It was actually five firms sharing resources with the same office. I just worked with one of them, Mr. Zublena. At the S.C.A.U, another partner Mr. Macary was more famous because he was excellent in public relations, which is why a lot of world famous architects worked with him: Tange, Pei, Koolhass, Botta, etc.. Being part of the office, I got indirectly to know these star architects.

Mr. Zublena was busy with business and administration. He was also a professor, who led me for my architect diploma project in France. So, he did not have much time to design by himself. The advantage was that he delegated the design projects to us. I first started to work in the company as a temporary employee, but the Tsinghua University's training made me capable to solve practical problems quickly and, after a month, I became an official employee. An American architect Tomas Sheehan worked with me, who arrived a few months earlier. He was in charge of the architectural issues, whilst I was working on planning. Tom was quite a genius; he could make beautiful sketches, free hand. His relaxed and artistic attitude had a strong influence on my work later. As a graduate from an engineering school, it is easy for me to be rigorous, but less easy to be imaginative. However, once you got that imagination, your rigor may help to realize it. We participated in two major competitions: one was "Technopole of Illkirsh in Strassburg"; the other was the "College of Mines in Nantes". We won both competitions and the two projects have been built.

H: From 1994, you worked at Foster's office in Hong Kong for three years. Foster is a high-tech orientated architect. In my observation, you got this orientation in your earlier designs. For instance, at the National Accounting Institute design, as well as the SongShan Lake Government Buildings project, where the use of glass and metal resulted in an exquisite building. Does that come from Foster? Or someone else, like Jean Nouvel?

Q: Norman Foster's "high-tech" reputation came probably from the H.S.B.C (bank of Hong Kong). That building made him a famous architect, and even led people to think that he could only do costly architecture. In fact, what you may not know, is that first, he never says "NO" to a client, because for him, the client's demand is the base of the design work; and on top of it, the project will always be carried out on time and within the budget. These are Foster's qualities. I remember once, he showed us an industrial project in Britain, in which the building was very banal, and the external wall was made just with simple corrugated sheets, but he was really proud of it, because the budget was so small that no one else could make this.

The H.S.B.C is extremely costly since the client wanted the most expensive office building in the world. At that time, after the signing of the Sino-British Declaration, the drain of capital was serious in Hong Kong, whilst the H.S.B.C held more than 90% of the capital in Hong Kong. In order to show their determination to stay and retain the capital, they had this precious building erected. As a precaution, they claimed a prefabricated and assembled building, which can be moved away elsewhere. If ever Hong Kong was really over? That is why you have a costly and high-tech design building there.

From the H.S.B.C project, Foster worked with industrial designers, some of them became his long-term partners. The word "high-tech" means at the same time "high" and "technology". Regarding "technology", people consider the functional pieces as useful as beautiful. From this point of view, whether gothic or Chinese traditional architecture, they all belong to that kind. However, "high" is relative to "low". Foster did not pursue high-tech throughout every project, or just showed the technology everywhere as a fundamental architectural expression. He tries to explore the technology, whether high or low, and being in coherence with them. For me, high-tech or not is not the issue, and what matters is that he treated each project as one single object. The Chek Lap Kok International Airport of Hong Kong was the largest project carried out in the office, and also the biggest terminal building in one phase all over the world at that time. Whilst, the smallest project carried out in the office at that time was just a door handle. Both were designed with a strong sense of unity.

Different from French architects, English architects proceed in a very logical way of design. Before starting on a design, they collect a lot of data, making analysis and inferences. In China, however, the imagination might be dissuaded by the restrictions from analysis and inferences. Well, whatever restrictions there may be, Foster will manage them and work them out in an unexpected and elegant design. Compared with other high-tech architects like Rogers, Foster's accentuates on elegance and unity. Piano's projects are integrated and elegant, similar to Foster's. Nowadays, when working on a project, I may break an entity first, but I will restore it for an integrated design. This comes from the Foster's influence.

I like buildings which are logical, inspired, elegant and generous.

H: Indeed, your design of Beijing National Accounting Institute gives an elegant, concise, stable and generous feeling.

Q: This project actually had a double effect on me: the facilities' part is more Foster's, but the residential part is probably more French.

H: In 1997, you participated in the design competition of Beijing Modern City; there were other competitors as Cui Kai and Zhu Xiaodi. You proposed the energy-saving concept and introduced the plants inside of the building. Then you were involved in some ecological planning projects. Sometimes you worked together with landscape architects. You showed therefore a certain concern for nature, energy saving and sustainability. Can you talk about this issue?

Q: "Green architecture" is everybody's concern. Norman Foster was the first architect who got a special financial fund for this kind of research in Europe. Therefore, he developed many relative technologies. Architects have to deal with many issues and, as long as he takes into consideration in his project the energy-saving and environmental protection, it is fine. The rest can be accomplished by professionals. Introducing green plants into a skyscraper building was just an idea. As that particular building was located on the south part of the street, consequently its northern façade faced to the street. Such a large elevation without sunshine was a pity. In order to reduce the northern façade to the least, I made a curved shape building. The sunlight can therefore reach almost every single part of the building, and that contributes to save energy. But this is just a planning story, nothing to do with the technology.

The intellectuals should always have a critical attitude. When people do not care about some important issues, you need to give them a warning. When everybody is aware of it, you may relax. At present, everyone talks about green architecture, energy saving, environmental protection and low carbon emission. I believe I have accomplished the mission and now leave it to others.

H: The Beijing National Accounting Institute is a high-tech oriented design, but later you changed your orientation. For instance, the Chuan Cheng project in Beijing and the Government Buildings in Songshan Lake, these two projects are more about the insertion of the buildings into their surroundings. After 2007, you paid more attention to the building's expression (or looks). More exactly, you made already different expressions through a range of façades in your Hepinglijing project in 2003. Another example was the windows arranged in an organic way in the User Friends Headquarters in 2004. What do you think of these changes?

Q: Like most world famous architecture firms, Foster's design seems to be a religion in his office. Architects at the office believed that only Foster's design is the best, and all the rest was just not worthy. It takes time to be out of Foster's influence, consciously or unconsciously. That change is natural, necessary and long.

H: It seems that you wanted to get out of it quickly.

Q: Foster is still a great architect for me. But, if you take him as God, you will be always under his shadow and remain his slave. You can never find yourself. Foster was a very important architect between 1980s and 1990s. When the 21st century comes, Herzog and Demeuron became the new stars who invested research with the building skin. Are they your model to follow? Nowadays, Zaha turns people's attentions to a free shape building. Shall we follow her?

Nobody is completely isolated, the surroundings will always be an influence. Does this means that you are not stable enough and it proves also that you are still alive. Otherwise you may always work in the same way without thinking. I have never been that confident in myself.

As architecture trends rise and fall, you realize that nothing is a hundred percent positive or negative. I do not want to follow trend, I am not interested in the past either. So I just try in my own way in different directions of design. That might not be easy, but at least that is fun.

H: Is that praising the new and rejecting the old?

Q: It is sometimes necessary if you undertake a creative profession. If you just want to sell dumplings, you can make the same dumplings and sell them day after day.

H: Judging by your recent buildings, I believe that you have fallen in love with building expressions. You made the square shape windows in various sizes for the Beckett factory in Beijing. In Wuhan, the façade is covered by a net skin in the Sales Office that plays with something in between the outside and the inside of a building; for the Software Park project in NanJing, you used the bamboo curtain as cladding system; for Tianjin Gulou's retail building, you put in the very thin columns arcade, and also for the Yuniaoliusu project in Hangzhou. All these buildings prove that your designs have changed from high-tech architecture into expression architecture. Please comment on this orientation?

Q: You may not be completely wrong, but neither correct. We are in a very superficial world in China today. When clients come to you, they are expecting something new and different. Suddenly, the building expression becomes a crucial issue. That is like forcing a maid to be a prostitute. Several years ago, the parents of my son's friend owned a piece of land in France. They asked me to design their house. I just made something very simple, put the windows and doors where it is necessary, with a courtyard in the middle, and that is it! They built it. As a building design it was just right. But in China, if ever you designed like this, people will think that you have done nothing, or that you have no talent.

However, each case should be treated correspondingly.

For the Tiancheng Sales Office in WuHan, the client demanded an attractive advertisement building in any style BUT Chinese. Actually, in WuHan, especially in HanKou where the project is located, most of the buildings belong to a colonial period with an eclectic style, which has nothing to do with China. A number of huge residential towers are being built next to the Sales Office. Therefore, only a relative big and special building can be noticed as an advertisement. The landscape design was already approved. They used crackled ice pattern for pavement. I superimposed that pattern on the elevation and provided shade. I wanted a hundred ladies to weave a gigantic basket with wickers, making it as the performance art, but it turned out to be in vain, and was then adopted by the Spanish Pavilion at the World Expo in Shanghai. In the end, the same pattern was made of prefabricated cast-in aluminum. Since there was a dockyard in Wuhan, it was easy for them to make it.

For the Beckett factory, it was another story. The client was American. Right from the beginning, I proposed a Chinese concept. He accepted it with pleasure when the scheme came out, because he really liked the change of architecture from the American style. So, expression is neither the only concern nor the major one for this project; the scheme tries to make an alive, local culture based space. Moreover, my effort is concentrated on solving the lighting issue of a huge internal area. The diagonal bracing member of the steel truss provides possibilities to make regular square sky windows, which diffuse a nice quality of light inside. They form a roof like a chessboard. This square pattern comes down gradually to the façade, avoiding the monotony of an industrial construction. The cube volume of building, the square courtyard, the windows and the chessboard roof work together to become an integrated design.

H: I still believe that you emphasize on architectural expressions. On the Qinhuai project, you made an abstract pattern on the building elevation which refers to the local historical towns; in the Xixi Clubhouse, as the patios provide the lighting and the ventilation, the external wall is free of those requirements, and you may therefore explore, without constraints, the relation between the building external wall and nature, such as mountains, trees or pools. In the Yuniaoliusu project, you extruded the forms, making the arcade as a ribbon. Meanwhile, both of them deal with the integration of fragments. What will you say of those two projects?

Q: You focus all the time on the façade. The big issue of the Qinhuai project is the planning influence rather than the façade. The original planning shows the straight streets and the huge squares. I was actually crashing the streets by buildings, making tensions between them for interesting events or scenarios. As a night activities area, unexpected stories should happen. With this design, the streets here are so narrow or so broad; likewise on the ground and in the air; and sometimes inside, sometimes outside of the buildings. That creates confusion. Under the influence of alcohol, people come across multiple opportunities to meet others, producing eventual love stories. That is the charm of a highly dense and crowded urban public space. The scheme tries to respond to the city, to the activities related and to the program, not just the façade.

Yuniaoliusu is a retail quarter, with a long pedestrian street and a square in the center. I was in charge of designing one side of the street. The other side of the street was to be designed by the architect Zhang Lei. Lei did not care about that particular street; he was just interested to make some internal streets in his plot. As my plot is in a bar shape, I had to deal with this. The project idea is making the architecture as a whole in keeping the specific identities of each store. The colonnade is actually the answer to the unity. You can hang the posters on it. The stores behind pillars twist their position each time, showing therefore their individuality. A continuous ridge ties up the all the stores, emphasizing again unity. It is an urban concept, different from the individual building design. In other words, making the individual buildings by means of urban design. It is definitely not about the façade.

The Xixi Clubhouse doesn't need to dialogue with its neighbor buildings, but with nature. I don't like to imitate nature. Whatever you build, you can not build more natural than nature itself, so it stays pseudo-nature. Instead, I prefer to deal directly with the geometry, which is very much artificial. The cladding used is made of some reflecting materials. The sky and the plants are reflected by the building and distorted into an abstract painting. Rhyming with the times and the seasons changing, the building responds to them continuously, sort of a simultaneous interpretation. Therefore, there is no conflict between human and nature: the people stay, so does nature.

H: The mirror effect is abstract and very much Chinese. Through the mirror, you can see yourself and the surroundings, which represents the reality and the past as well. Both are fused into one image. It is just like obscurity from tissue paper, the vagueness of a bronze mirror, the transparence and opaqueness of jade, the gradations from clear ink to dark, from a thin line and thick dot in painting and calligraphy. The vagueness and obscurity reflects the culture and its signification. Is that an expression of Chinese mind?

Q: I haven't thought that much. If there is something Chinese, it comes from the program. Actually, in the Ming and Qing dynasties, quite a lot of poets and writers built houses there, leaving their poems and paintings. For that reason, people built some old style buildings at the first and second phases of construction in Xixi. Our job is about the third phase of Xixi construction. I refused to memorize only one period of history. When I first went to the site, I saw the farmers' houses there, which was very interesting. People mixed the clear and dark ceramic tiles on the wall. They like to use the stainless steel as material for handrail and protection grills, and put a huge stainless steel ball on the top of the building.

History is a long river. You can't say which period is more valuable. I granted equal dignity and quality to ancestors of different periods of the past. For tracing them, their similarity is what I am interested in. We noticed that either people from the Ming and Qing dynasties or of our time make houses with a simple geometry; the roof is always sloping; the complexity comes from the interweaving and overlapping of a simple unit. So we defined our basic unit as twelve by twelve meters square. We combined them together freely with always a slopping roof. That results in a very various and organic geometry. I just design one elevation with plenty of holes, and put it over on every single façade. If the façade falls where there is no need of openings, I just erase the hole off.

The geometry belongs to the Ming dynasty as to the People's Republic of China. It belongs to the past as to the present and to the future. It belongs to China as to the world. I didn't deliberately celebrate China.

As for the mirror, what impressed me most were the Amsterdam's canals. The buildings, the lights, the trees and people are distorted in water with reflection. I used the concrete objects and turned them into an illusionary image, letting the world less realistic.

H: Let's talk about the architecture. Do you have some particular design philosophy?

Q: No. That is the concern of the media. Each firm has to work out its philosophy for itself. When you look at those philosophies, they are all correct. Some make a point on society or politics or culture, some on city or space or geometry, or on construction or technology or material, or on lighting effect, or color, or sustainability, or service, or people relationship. You put all these together, that is my philosophy.

H: What is your singularity?

Q: You look at the work of architects like Ando, Gehry or Zaha, you know what they do and what they are going to do. I don't like to be frozen. I want to go wherever I like to. You have put Jean Nouvel in the category of high tech architects like Foster. However, Nouvel is keep-changing his architectural language, and you never know what he is going to do next! I prefer that.

H: You do not want to limit yourself by one style. But people can still recognize your design. Your drawings are also very special. Though you refuse to define a style, people can still feel your style.

Q: Xuan Zang (a famous Chinese monk who brought Buddhism from India to China) drew a circle, so that you can never get out of this circle. That circle is the palm of Buddha. That is your environment, your experience and culture. You can never jump out of the palm of Buddha.

H: Is this your philosophy without definition?

Q: May be...

H: Why can't you trace your own way and have a personal Qi Xin's architectural style?

Q: I have always been very much on my own way. Every step has been accomplished entirely by my self. Actually, I am not interested in creating a trade mark. Instead, I prefer working in a natural way, which means that only the specific situation of each project will be the guide of the inspiration for a singular and appropriate architecture. Therefore, I do not know where my next step will take.

H: Last question: when people write the history, they put architects of a certain period into different schools. You could not find your seat if you do not make a reservation. Or you are expecting historians to make their story in another way?

Q: I think that few architects like being catalogued, me neither. So, let historians do their work on their side and let architects work on theirs.

摄影：曹扬

团队

齐欣建筑设计咨询有限公司由齐欣、秦岩、贺炜、张江及杨军创建于2002年。从事与建筑相关的（包括城市设计、景观、室内等）建筑、结构、设备设计咨询服务。现有成员：戴伯军、高银坤、贺炜、刘阳、罗彬、齐欣、秦岩、徐丹、于向东、袁维、张江、张亚娟、赵胜涛。

Team

Qi Xin Architects and Engineers was founded in 2002 by Qi Xin, Qin Yan, He Wei, Zhang Jiang and Yang Jun. The scope of works covers architecture, structure and M&E design services and documentation on city planning, architecture, landscape and interior arrangement. The actual staff includes: Dai Bojun, Gao Yinkun, He Wei, Liu Yang, Luo Bin, Qi Xin, Qin Yan, Xu Dan, Yu Xiangdong, Yuan Wei, Zhang Jiang, Zhang Yajuan, Zhao Shengtao.

其他作品年表
Chronology of other Works

●——实现作品　○——未实现作品

名称：城市别墅 ●
位置：天津
功能：会所
规模：3000m²
甲方：亚资置业
设计：维思平
时间：2001～2002
阶段：竣工

名称：北京会议中心 ●
位置：北京
功能：会议室
规模：350m²
甲方：北京会议中心
设计：齐欣建筑 + 联安国际设计公司
时间：2002
阶段：竣工

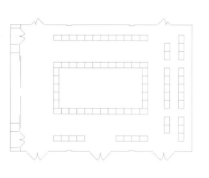 

名称：天科大厦 ●
位置：北京
功能：办公
规模：67500m²
甲方：天创房地产开发公司
设计：齐欣建筑 + 联安国际设计公司
时间：2002～2003
阶段：竣工

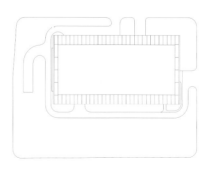 

名称：松山湖实验剧场 ○
位置：东莞
功能：剧场
规模：8000m²
甲方：松山湖管委会
设计：齐欣建筑 + 深圳华森设计公司
时间：2002
阶段：方案

名称：和平丽景 ○
位置：廊坊
功能：住宅 + 商业
规模：2300m²
甲方：新奥置业
设计：齐欣建筑
时间：2003
阶段：概念

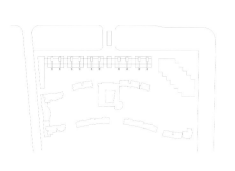

名称：齐欣事务所 ●
位置：北京
功能：办公
规模：400m²
甲方：齐欣建筑
设计：齐欣建筑
时间：2003
阶段：竣工

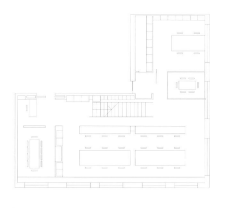

名称：金陵神学院 ○
位置：南京
功能：学校
规模：30000m²
甲方：金陵神学院
设计：齐欣建筑
时间：2004
阶段：竞赛

名称：用友软件园 ○
位置：北京
功能：办公
规模：35000m²
甲方：用友软件
设计：齐欣建筑
时间：2004
阶段：方案

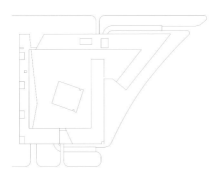

名称：香山传城住宅区 ○
位置：北京
功能：住宅
规模：75000m²
甲方：兴荣基房地产
设计：齐欣建筑
时间：2004
阶段：概念

名称：顺驰总部装修 ●
位置：北京
功能：办公
规模：2400m²
甲方：顺驰中国控股
设计：齐欣建筑 + 建学建筑设计公司
时间：2005
阶段：竣工

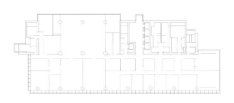

名称：中国京剧院 ○
位置：北京
功能：办公 + 剧场
规模：
甲方：光大房地产
设计：齐欣建筑
时间：2005
阶段：概念

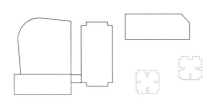

名称：半山公寓 ○
位置：深圳
功能：住宅 + 酒店
规模：18000m²
甲方：招商地产
设计：齐欣建筑
时间：2005
阶段：概念

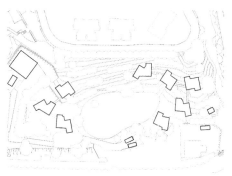

名称：观景阁 + 敦洲坊 ○
位置：深圳
功能：景观 + 酒吧
规模：
甲方：深圳规划局滨海分局
设计：齐欣建筑
时间：2005
阶段：概念

名称：反正清华 ○
位置：北京
功能：办公
规模：1500m²
甲方：清华科技园
设计：齐欣建筑
时间：2005
阶段：竞赛

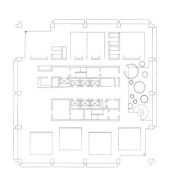

名称：盛木·天一方 ●
位置：海口
功能：住宅
规模：42000m²
甲方：美丽沙房地产
设计：齐欣建筑 + 海南华磊设计公司
时间：2005
阶段：竣工

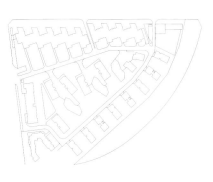

名称：犹太学校装修 ●
位置：北京
功能：幼儿园
规模：700m²
甲方：犹太学校
设计：齐欣建筑 + 建学建筑设计公司
时间：2005
阶段：竣工

名称：鼓楼区商业街 ●
位置：天津
功能：商铺
规模：3000m
甲方：中新集团产
设计：齐欣建筑 + 天津市房屋鉴定设计院
时间：2005～2008
阶段：竣工

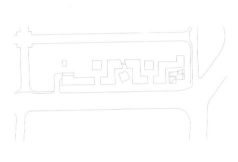

名称：玉河街区改造 ○
位置：北京
功能：商业 + 酒店
规模：3500m²
甲方：时尚生活
设计：齐欣建筑
时间：2006
阶段：概念

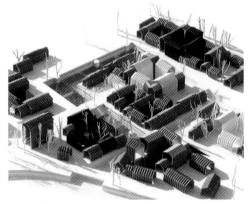

名称：金地售楼中心（幼儿园）●
位置：北京
功能：音体教室
规模：800m²
甲方：金地地产
设计：齐欣建筑 + 北京新纪元设计公司
时间：2006
阶段：竣工

名称：贝克特厂房 ●
位置：北京
功能：厂房
规模：4500m²
甲方：贝克特太平洋
设计：齐欣建筑 + 北京中铁工设计院
时间：2006～2008
阶段：竣工

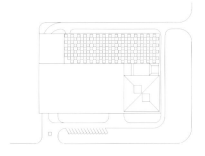

名称：歌华开元大酒店 ●
位置：北京
功能：酒店
规模：立面设计
甲方：歌华传媒
设计：齐欣建筑 + 北京市建筑设计研究院
时间：2006～2008
阶段：竣工

名称：Town 中堂售楼中心 ●
位置：天津
功能：会所（售楼处）
规模：1200m²
甲方：中新集团
设计：齐欣建筑 + 墨臣事务所
时间：2006～2008
阶段：竣工

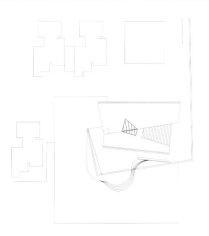

名称：秦淮风情街 ○
位置：南京
功能：商业 + 酒吧
规模：18000m²
甲方：江宁区东山街道政府
设计：齐欣建筑
时间：2007
阶段：施工图

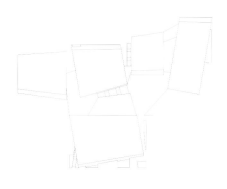

名称：中海天津 ○
位置：天津
功能：办公 + 酒店
规模：40000m²
甲方：中海地产
设计：齐欣建筑
时间：2007
阶段：方案

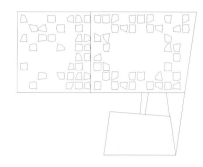

名称：奥运树信息柱 ○
位置：北京
功能：信息 + 照明
规模：
甲方：中奥集团
设计：齐欣建筑
时间：2007
阶段：概念

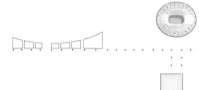

名称：上实湖州办公楼 ○
位置：湖州
功能：办公
规模：12000m²
甲方：上实集团
设计：齐欣建筑
时间：2008
阶段：概念

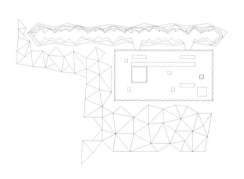

名称：生态绿谷综合体 ○
位置：重庆
功能：商业
规模：76000m²
甲方：阳光100房地产
设计：齐欣建筑
时间：2008
阶段：概念

名称：泰达研发社区 ●
位置：青城山
功能：办公
规模：70000m²
甲方：成都泰达
设计：齐欣建筑 + 标鼎时代
时间：2008 至今
阶段：施工

名称：泰达研发社区酒店 ○
位置：青城山
功能：酒店
规模：9000m²
甲方：成都泰达
设计：齐欣建筑
时间：2008
阶段：方案

名称：商业综合体 ○
位置：南京
功能：商业 + 酒店
规模：70000m²
甲方：泰合地产
设计：齐欣建筑
时间：2008
阶段：方案

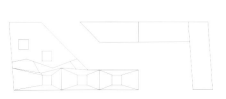

名称：万庄生态城 ○
位置：廊坊
功能：居住 + 办公 + 酒店
规模：55hm²
甲方：上实集团
设计：齐欣建筑
时间：2008
阶段：概念

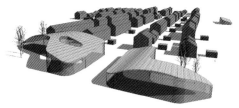

名称：比克大厦 ●
位置：天津
功能：办公
规模：120000m²
甲方：比克电池
设计：齐欣建筑 + 郑州大学设计院
时间：2008 至今
阶段：施工

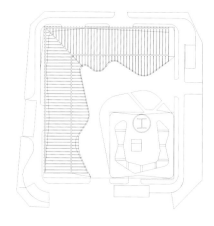 

名称：广陵新城城建展览馆 ○
位置：扬州
功能：展览
规模：6000m²
甲方：南京新城
设计：齐欣建筑
时间：2008
阶段：概念

名称：民乐村重建 ○
位置：绵竹
功能：民居
规模：500 户
甲方：志愿设计
设计：齐欣建筑
时间：2008
阶段：概念

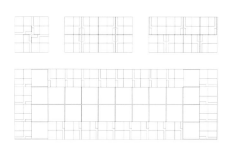 

名称：武清演艺中心 ○
位置：天津
功能：剧场 + 影院
规模：16000m²
甲方：武清规划局
设计：齐欣建筑 + 北京市建筑设计研究院
时间：2009 至今
阶段：施工图

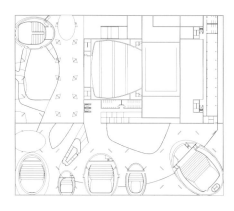 

名称：唐山党校 ○
位置：唐山
功能：学校
规模：76000m²
甲方：唐山市委
设计：齐欣建筑
时间：2009
阶段：竞赛

 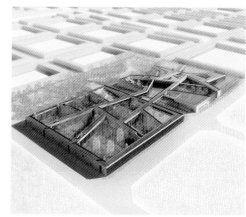

名称：生态示范居住区 ○
位置：唐山
功能：住宅
规模：403000m²
甲方：万年基业
设计：齐欣建筑
时间：2009
阶段：竞赛

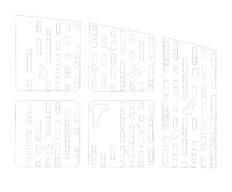 

名称：渤龙湖商业街 ●
位置：天津
功能：商业 + 酒店
规模：46600m²
甲方：海泰集团
设计：齐欣建筑 + 天津华汇设计有限公司
时间：2009 至今
阶段：施工

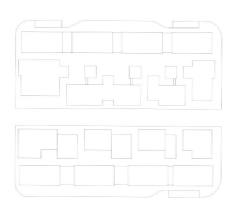 

名称：渤龙湖住宅区 ●
位置：天津
功能：住宅
规模：107000m²
甲方：海泰集团
设计：齐欣建筑 + 清华建筑设计研究院
时间：2009 至今
阶段：施工

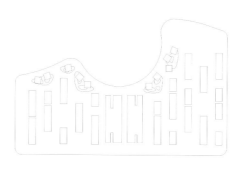 

名称：综合楼 ○
位置：北京
功能：教学
规模：72000m²
甲方：中国石油大学
设计：齐欣建筑
时间：2010
阶段：竞赛

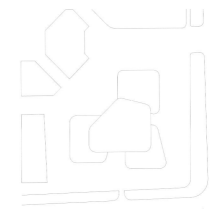 

名称：塔什干文化中心 ○
位置：乌兹别克斯坦
功能：展览 + 餐饮
规模：2600m²
甲方：匿名
设计：齐欣建筑
时间：2010
阶段：方案

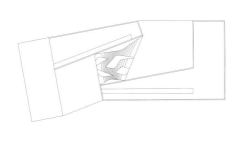 

名称：神农坛 ○
位置：株洲
功能：祭拜
规模：景观场所
甲方：株洲市委
设计：齐欣建筑
时间：2010
阶段：概念

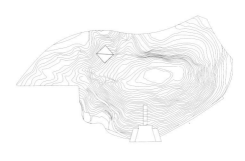 

名称：丰台科技园东区三期规划 ○
位置：北京
功能：办公 + 酒店
规模：300000m²
甲方：中铁华生
设计：齐欣建筑
时间：2010
阶段：概念

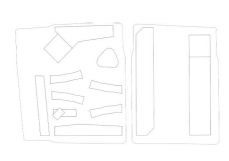 

名称：东区音乐公园 32 号楼改造 ○
位置：成都
功能：歌舞厅 + 培训
规模：19000m²
甲方：成都传媒文化
设计：齐欣建筑
时间：2010
阶段：方案

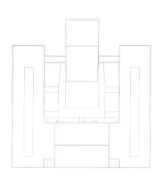 

名称：20+10 办公 ●
位置：鄂尔多斯
功能：办公
规模：18000m²
甲方：鄂尔多斯东胜区区政府
设计：齐欣建筑 + 互联盟
时间：2010 至今
阶段：施工图

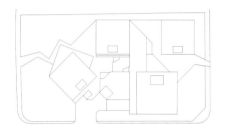 

名称：运河水乡规划 ○
位置：北京
功能：商业 + 餐饮 + 酒店
规模：140000m²
甲方：海航集团
设计：齐欣建筑
时间：2010
阶段：概念

名称：国际贸易服务中心 ○
位置：北京
功能：展览 + 销售 + 办公
规模：42000m²
甲方：歌华传媒
设计：齐欣建筑
时间：2010
阶段：概念

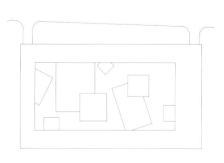 

名称：康巴什独立住宅群 ○
位置：鄂尔多斯
功能：住宅
规模：17500m²
甲方：互联盟
设计：齐欣建筑
时间：2011
阶段：概念

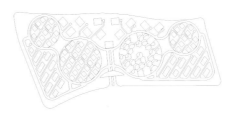 

名称：昆山北站交通枢纽 ○
位置：昆明
功能：办公 + 公交枢纽
规模：69000m²
甲方：昆明轨道
设计：齐欣建筑 + 铁四院
时间：2011 至今
阶段：扩初

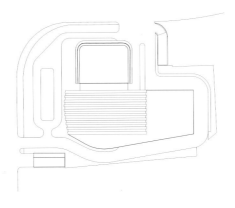 

名称：南大街改造 ○
位置：安阳
功能：商业 + 办公
规模：350000m²
甲方：北建工
设计：齐欣建筑
时间：2011
阶段：概念

名称：就业局办公楼 ○
位置：鄂尔多斯
功能：办公 + 宿舍
规模：42500m²
甲方：就业局
设计：齐欣建筑 + 北建工设计院
时间：2011 至今
阶段：施工图

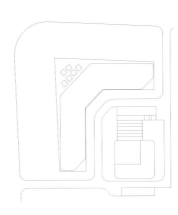 

名称：空客培训中心大堂 ○
位置：北京
功能：接待 + 休息
规模：400m²
甲方：空客北京培训中心
设计：齐欣建筑
时间：2011 至今
阶段：施工图

名称：木兰围场庄园 ○
位置：承德
功能：会所 + 酒店
规模：3000m²
甲方：力保北方
设计：齐欣建筑
时间：2011 至今
阶段：方案

名称：市民活动中心 ○
位置：乌兰察布
功能：文体娱乐
规模：27000m²
甲方：乌兰察布市委
设计：齐欣建筑
时间：2011 至今
阶段：概念

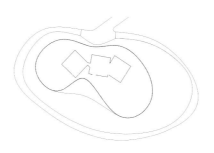

名称：苇沟会所 ○
位置：北京
功能：会所
规模：6000m²
甲方：诚通嘉业
设计：齐欣建筑
时间：2011 至今
阶段：方案

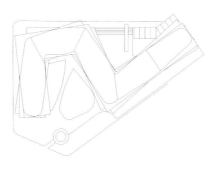

齐欣简介

清华大学建筑系学士文凭；
巴黎贝勒维尔建筑学院研究生文凭；
巴黎拉维莱特建筑学院建筑师文凭。

1990～1993年任法国巴黎建筑与城市规划设计院建筑师；
1994～1997年任英国福斯特亚洲建筑设计事务所高级建筑师；
1997年被清华大学特聘为外国专家、副教授；
1998～2001年任京澳凯芬斯设计有限公司总设计师；
2001～2002年任德国维思平建筑设计咨询有限公司总设计师；
自2002年起任北京齐欣建筑设计咨询有限公司董事长兼总建筑师。

2002年获WA建筑奖；
2004年获亚洲建筑推动奖；
2004年获法国文化部授予的艺术与文学骑士勋章；
2010年获全国优秀工程勘察设计行业一等奖；
2011年获北京国际设计三年展建筑设计奖。

近年来先后应邀参加了法国蓬皮杜中心"中国当代艺术"展，法国建筑博物馆中国当代建筑师"立场"展，北京歌华艺术中心齐欣个展，深圳第一、第二及第三届城市与建筑艺术双年展，罗马"向东方"中国城市与建筑展。

时任中国建筑学会会员、中国扶贫基金会灾后重建专家委员会专家、北京昌平区城乡规划与公共艺术委员会专家委员会专家、天津大学新校区专家委员会专家、乌兰察布市规划建设评审咨询专家委员会专家、及《世界建筑》，及《城市·环境·设计》和《建筑细部》杂志社编委。

Profile

Bachelor, Department of Architecture, Tsinghua University, Beijing;
CEAA, School of Architecture, Paris-Belleville;
DPLG, School of Architecture, Paris-La Villette.

Architect at SCAU, between 1990 and 1993 in Paris;
Senior architect at Norman Foster's office between 1994 and 1997 in Hong Kong;
Expert and associate professor at Tsinghua university in 1997;
Chief designer at Jing-Ao Kann Finch between 1998 and 2001;
Chief designer at WSP between 2001 and 2002;
Qi Xin Architects and engineers President and chief architect since 2002.

WA award in 2002;
Architecture Promoting Award of Asia in 2004;
Chevalier des Arts et des Letters in 2004;
First prize of excellent design of China in 2010;
Architecture Design Award of Beijing International Triennium Design Exhibition in 2011.

In recent years Qi Xin has been invited to "Alore la Chine"exhibition in Pompidou Center in Paris, "Position"exhibition at Architecture museum in Paris, Qi Xin 10 years exhibition in GeHua art center in Beijing, 1st, 2nd and 3rd ShenZhen Architectural Biennale, "Vesso Est"exhibition in Rome.

Member of China Architects Association, Expert of Post-disaster Reconstruction Advisory Committee, of China Foundation of Poverty Alleviation, Expert of Advisory Committee of Urban Planning, and Public Art Committee of Beijing Changping District, Expert of Advisory Committee of Tianjin University New Campus, Expert of Advisory for Planning and Construction Committee, of Wulanchabu City, Editorial Board member of "World Architecture". "Urban Environment and Design"and"Detail"magazines.

致 谢

感谢王昀建筑师倡议本套丛书的出版；感谢中国建筑工业出版社和金晶科技公司对本图书的投入和推动；感谢单戴娜、蓬莓及 Robert Backer、Jochen Steppich、Andreas Stadlinger 和 Marie Josèphe Dumont-Qi 对图书文稿的梳理和纠正；感谢付兴、方振宁、杨超英、舒赫、姚力和曹扬的精心摄影；感谢这些年来业主对我的信任与支持和业内及各界朋友们的帮助和关爱；感谢事务所秦岩、贺炜、张江十年来的通力合作，以及所有曾经或依然在事务所工作的同事们的努力。

Acknowledgement

Thanks to Architect Wang Yun who initiates this bookseries; thanks to China Architecture & Building Press and the Shandong Jinjing Science & Technology Ltd. who edit and support this book; thanks to Diana Chan-Chieng, Peng Mei, Robert Backer, Jochen Steppich, Andreas Stadlinger and Marie Josèphe Dumont-Qi for their translations and corrections of the texts; thanks to Fu Xing, Fang ZhenNing, Yang Chaoying, Shu He, Yao Li and Cao Yang for their photographs; thanks to the clients for their trust; thanks to all the friends who gave me the supports through all those years; thanks to Qin Yan, He Wei and Zhang Jiang for their fully cooperation during the past 10 years and even longer; thanks to all the colleagues of the past and the present for their efforts.